AF543270

FUNDAMENTALS OF STRUCTURAL DESIGN

FUNDAMENTALS OF STRUCTURAL DESIGN

BY

LIAM ANDERSON
CES REYNOLDS

KNOWLEDGE
BAKERS

KNOWLEDGE BAKERS

© 2024 AUTHORS

ISBN: 9789390013371 (HB)

All rights reserved under International Copyright Conventions. No part of the book or part thereof, including title of the book, be reproduced in any printed form or language without written permission of the author and the publishers.

Published by : **Knowledge Bakers**
Pune, India
Email: knowledgebakers2019@gmail.com

Printed at : **Replika Press Pvt Ltd**

PREFACE

Structural design is the methodical investigation of the stability, strength and rigidity of structures. The basic objective in structural analysis and design is to produce a structure capable of resisting all applied loads without failure during its intended life. The primary purpose of a structure is to transmit or support loads. If the structure is improperly designed or fabricated, or if the actual applied loads exceed the design specifications, the device will probably fail to perform its intended function, with possible serious consequences. A well engineered structure greatly minimizes the possibility of costly failures. Structural designing provide all the crucial information about foundations, walls, floors, roof types, steep beams, material quality, and more to ensure that the structure is built to be safe and sound. It is essential for civil engineers to work with design aesthetics and build structures that can withstand its loads and pressures without leading to any mishaps. Indeed, building a structure without considering the structural design of it always remains under the high risk of collapsing under its own weight or due to the environmental pressures. Structural designing is hence crucial to construction projects as it directly impacts the safety and durability of the structures.

Civil engineers are required to perform a thorough and detailed analysis of the structure with respect to the principles of the structural design. This helps to ensure that the new structure meets up all the necessary safety criteria and design codes. Moreover, when an in-depth analysis of the structure is made, it helps bring out information about the load that the building is capable of bearing, the wind velocity it can withstand, and the overall utility and capability of the building under different environmental conditions. For instance, knowing the accurate estimation of the magnitude of tremors a skyscraper can withstand and the safe weight maximums on bridges can surely help prevent any catastrophes in future.

The book provides an in-depth understanding of different types of loads that act on structures, including dead loads, live loads, wind loads, earthquake loads, and snow loads. It also covers how to calculate these loads and their effects on structures. The book emphasizes the importance of equilibrium in structural design and provides practical examples to help readers understand the concept. The book covers specialized structural elements such as catenaries cables, arches, trusses, and lattice domes. It also covers how to determine the strength and durability of these materials. The book covers how to calculate and analyze shear and bending stress in simple beams, which are fundamental structural elements and explains how to calculate and analyze deflection in simple beams and its effects on structural design. The book covers load tracing, which is a method used to understand how loads are transferred through a structure. The book is highly useful for students, teachers, researchers and professionals in the field of civil engineering and structural design.

We are grateful to all those persons as well as various books, manuals, periodicals, magazines, journals etc. that helped in the preparation of this book. In spite of the best efforts, it is possible that some errors may have occurred into the compilation and editing of the book. Further queries, constructive suggestions and criticisms for the improvement of the book are always welcomed and shall be thankfully acknowledged.

Liam Anderson
Ces Reynolds

CONTENTS

PART I

MAIN BODY

CHAPTER 1

Loads on Structures

This chapter will discuss different types of loads and how you may consider them in structural design. Before analyzing the statics and mechanics of structures, it is essential to learn how much load a structure may carry.

DIFFERENT TYPES OF LOADS ON STRUCTURES

• Static loads

"Static loads are assumed to be applied slowly to a structure until it reaches its peak value without fluctuating rapidly in magnitude or position. Under a static load, a structure responds slowly, and its deformation reaches a peak when the static force is maximum." [1]

• Dynamic loads

"Dynamic loads are applied suddenly to a structure, often with rapid changes in magnitude and point of application. Under a dynamic load, a structure develops inertial forces in relation to its mass, and its maximum deformation does not necessarily correspond to the maximum magnitude of the applied force. The two major types of dynamic loads are wind loads and earthquake loads". [1]

• Impact loads

An impact load, which is categorized as a dynamic load in some references, acts rapidly on a structural system and causes vibration. The created vibration hinders the structural system from reaching the state of equilibrium.

Loads on structures are classified into the following categories:

- Dead load – D
- Floor live load – L
- Roof live load – L_r
- Snow load – S
- Wind load – W
- Earthquake loads – E
- Rain load – R
- Soil and Hydrostatic Pressure and Flood Loads – H
- Ice load

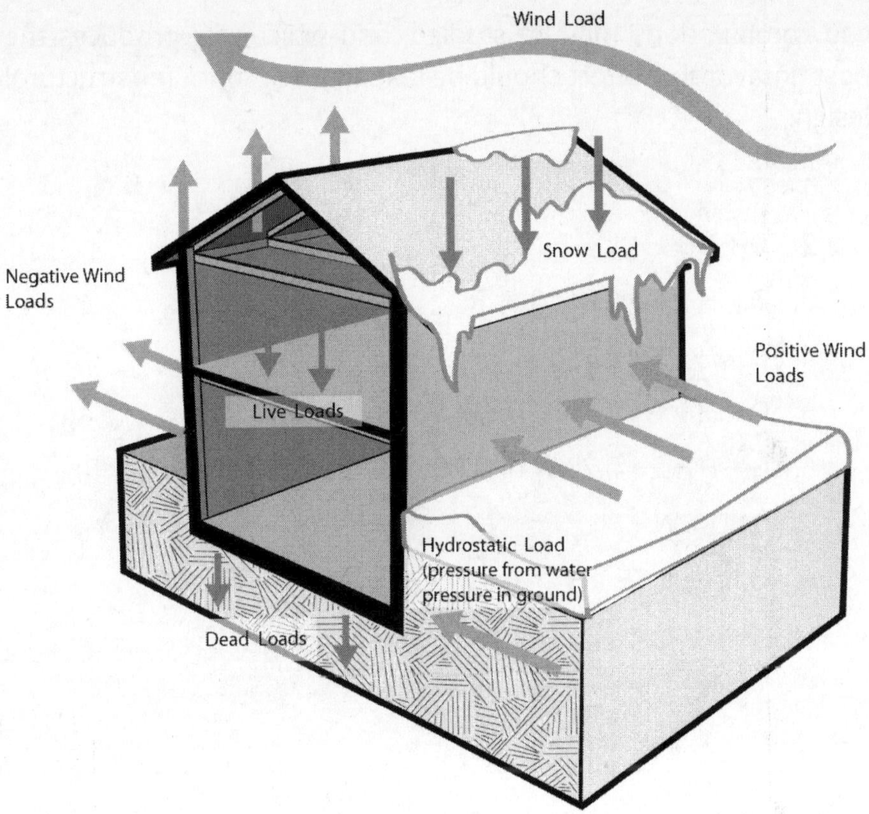

Figure 1-1: Some types of loads that act on a typical building

The American Society of Civil Engineers (ASCE) provides a standard, SEI/ASCE 7-02, that addresses the definition of different types of loads and determines minimum design loads for buildings.

Combinations of loads

Buildings are not designed for ALL the possible loads that may affect them, but rather a reasonable combination of loads. Buildings will rarely be impacted by the dead load, maximum of live load, snow load, earthquake load, wind load, and an impact load at a single moment. The reasonable combinations of the loads are also defined in the ASCE 7 Standard. For example, the following load combinations may be studied, and whichever produces the most unfavorable effect should be taken into account for structural design.

1. D+ F
2. D + H + F + L + T
3. D + H + F + (L_r or S or R)
4. D + H + F + 0.7S(L + T) + 0.7S(L_r or S or R)
5. D + H + F + (W or 0.7 E)
6. D + H + F + 0.7S(W or 0.7E) + 0.7SL + O.7S(L_r or S or R)
7. 0.6D + W + H
8. O.6D + 0.7E + H

Dead load

Dead load is simply the self-weight of all major construction materials, including the building structures, permanent walls, fixed utilities, and equipment (e.g., HVAC systems). The weight of each structural component can be measured by multiplying the volume of components by their density (Mass = Volume x Density).

Density (pounds per cubic foot)=(Mass (lbs))/(Volume (cubic feet))

Dead load (lbs)= Density (pcf) x Volume (cf)

Dead load units:

Pounds per linear foot (plf) → plf is used for describing the DL of beams and columns

Pounds per square foot (psf) → psf is used for describing the DL of slabs

For example, the weight of the wood beam shown below can be computed and described in pounds per linear foot (plf) by multiplying its density of 35 pounds per cubic foot (PCF) and its volume.

DL = volume x density

(11.25/12 ft) X (3.5/12 ft) x 1 ft x 35 pcf = 9.57 plf

(3.5)/12 x (11.25)/12 ft^2 x 35 lbs = 9.57 plf

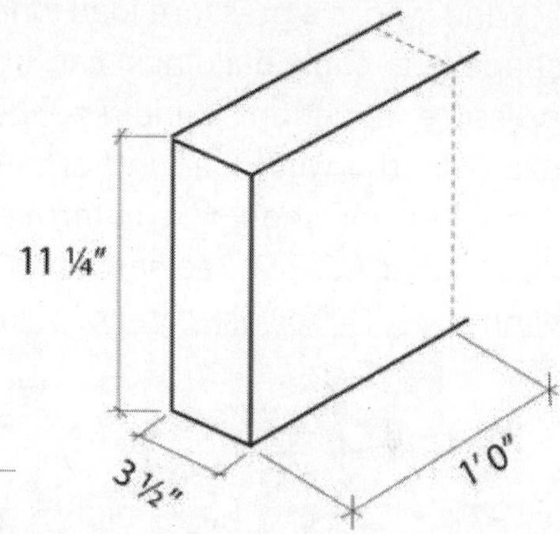

Live load

Live load is based on occupancy and includes the weight of the people, furniture, machinery (e.g. in a factory), vehicles, and equipment being used temporarily. There are two types of live loads:

- Floor live load
- Roof live load (e.g. the weight of roof garden or the maintenance personnel and their equipment)

The live loads used in the structural analysis should not be less than the minimum uniformly distributed unit loads required by building codes. You may refer to the International Code Council (ICC) to learn more about the estimated live loads in different spaces for various occupancies (see table of minimum uniformly distributed live loads and minimum concentrated live loads).

Wind load

The wind blowing against a building may be strong and cause discomfort or may be extreme, such as a tornado, hurricane, or heavy storm, and cause destruction. While wind load, like earthquake load, is three-dimensional in nature, its horizontal component is considered more critical in structural design. Thus, wind load is classified as a lateral load.

Wind load is a pressure load whose effect on buildings of regular shape (e.g. cubic buildings) can be analyzed considering the basic pressure equation. Wind speeds vary in different zones. For example, the wind load in Portland, OR, is 16 psf. Wind loads on every building should be determined in accordance with Chapters 26 to 30 of ACSE 7. See section 1609 and the map of basic design wind speed, V, for risk category II, III, and other structures.

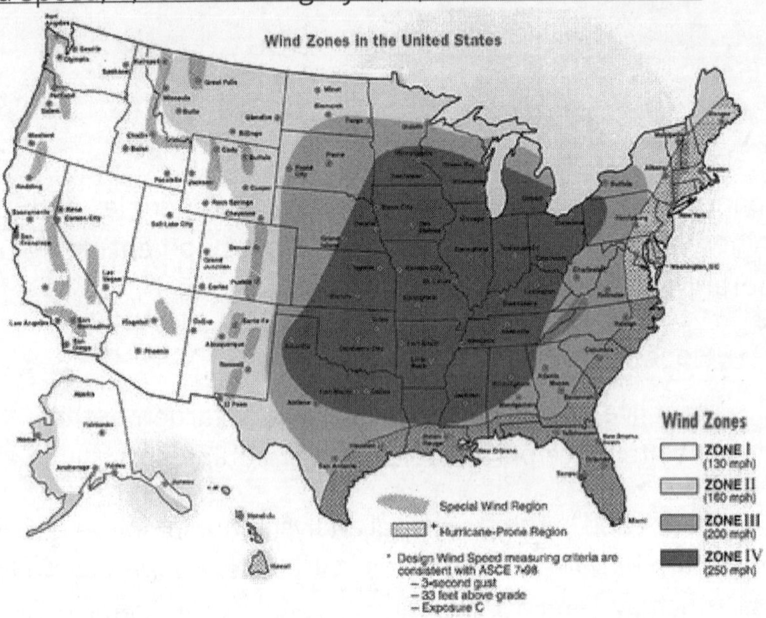

Figure 1-2: Wind zones in the United States of America (Source: USA Natural Hazards Map, 2012, https://www.flickr.com/photos/konabish/6810939678)

If the configuration of a building is complex or it has a high importance factor, you may need to conduct a wind tunnel test. There are not many laboratories around that do the wind tunnel test because it is expensive to have the facilities. You can send your design to those labs, and they will send back the pressure data to you. Then, engineers can design a suitable structure. Watch video 1-1 (https://www.youtube.com/watch?v=UEgk2Bgz16s&t=1s)

to see flow visualization around simple building shapes in a wind tunnel. In the following, video 1-2 (https://www.youtube.com/watch?v=tHMPR7flpf4&t=250s) shows how tall buildings tame the wind.

One or more interactive elements has been excluded from this version of the text. You can view them online here: https://pdx.pressbooks.pub/archistructures/?p=5#oembed-1

Video 1-1: Flow visualization around simple building shapes in the wind tunnel

One or more interactive elements has been excluded from this version of the text. You can view them online here: https://pdx.pressbooks.pub/archistructures/?p=5#oembed-2

*Video 1-2: How tall buildings tame the wind*Some software systems allow you to simulate the test. WIND CFD (Computational Fluid Dynamics) software is one of them. You can use the results of these simulation programs in the conceptual phases of design. But for critical buildings such as towers, the outcome of the simulation programs is not sufficien for specifying the structural system, and a wind tunnel test is still required.

Snow load

Snow loads result from the weight of snow accumulating on a flat or sloped roof. Snow loads vary with geographic location, wind condition, geometry and slope of the roof, building, and site exposure. For example, a minimum of 54 psf snow load should be considered for buildings constructed in Montréal, QC, while 10 psf snow load is recommended for designing structures in Portland, OR. See section 1608 and the map of ground snow loads for the United States.

Earthquake load

The tectonic plates are always slowly moving, but they get stuck at their edges due to friction. When the stress on the edge overcomes the friction, an earthquake releases energy in waves that travel through the earth's crust and causes the shaking that we feel. An earthquake is caused by a sudden slip on a fault and may consist of a series of longitudinal and/or transverse waves known as P-wave or S-wave, respectively. A P-wave is the fastest wave.

Although earthquake vibrations are three-dimensional, their horizontal components are more critical in designing structural systems. Thus, generally, we consider earthquake load as a lateral force acting on a building. The earthquake load depends on how close the building is to fault lines. To read more about earthquakes, see "What is an earthquake and what causes them to happen?". You may explore the latest earthquakes map and list for the U.S. and worldwide.

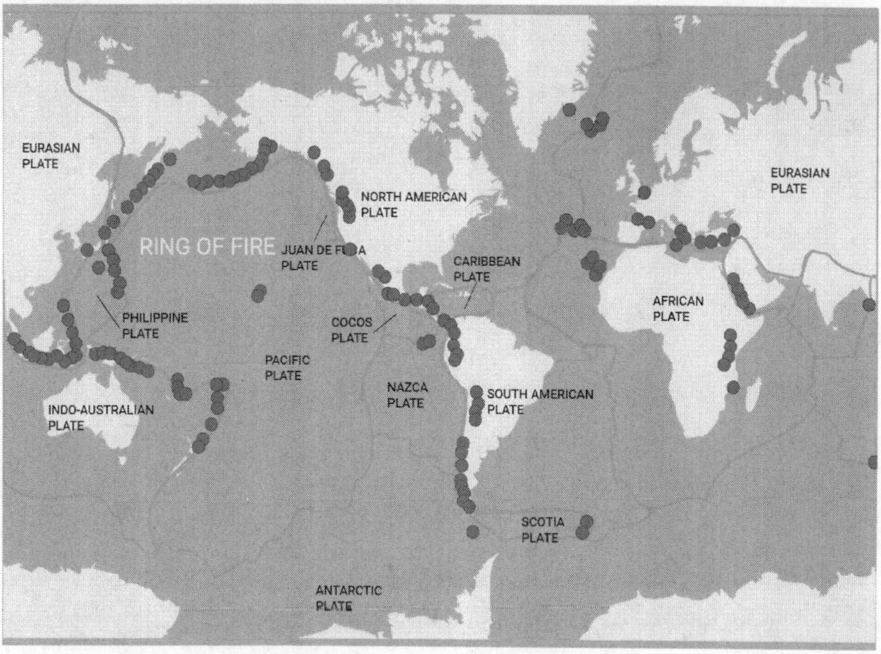

Figure 1-3: Tectonic plates

In addition to various types of damage to building structures, there are three major ground failures caused by earthquakes:

- Landslides, particularly when the land is wet.
- Liquefaction: Sandy soils can be liquified by even a light ground shaking when saturated by water.
- Subsidence: Non-uniform ground sink

In the following, video 1-3 (https://www.youtube.com/watch?v=e7ho6z32yyo&t=1s) explains how earthquakes impact buildings and our cities. Video 1-4 (https://www.youtube.com/watch?v=jhRuUoTnA6g&t=2s) discusses if the earthquakes are predictable. Video 1-5 (https://www.youtube.com/watch?v=H4VQul_SmCg&t=2s) shows how earthquakes impact tall buildings.

One or more interactive elements has been excluded from this version of the text. You can view them online here: https://pdx.pressbooks.pub/archistructures/?p=5#oembed-3

Video 1-3: How earthquakes impact buildings and our cities.

One or more interactive elements has been excluded from this version of the text. You can view them online here: https://pdx.pressbooks.pub/archistructures/?p=5#oembed-4

Video 1-4: Are earthquakes predictable?

One or more interactive elements has been excluded from this version of the text. You can view them online here: https://pdx.pressbooks.pub/archistructures/?p=5#oembed-5

*Video 1-5: How buildings fall in ear*thquakes.

TOPICS FOR CRITICAL THINKING

- Considering the reality of climate change, we have been experiencing unexpected extreme weather conditions, such as wildfire in mild and humid regions, floods in ari areas, and snowstorms in semi-arid or marine west coast regions. How do current load requirements for structural design respond to unexpected weather conditions? Should we reconsider load requirements in building codes?
- Should we treat all buildings the same? For example, should the structural design requirements be the same for hospitals, fire or police departments, and warehouses?
- How does a tuned mass damper support building resiliency against earthquakes?
- How does a base isolation system support building resiliency against earthquakes?

CHAPTER 2

Forces and Vector Analysis

REPRESENTATION OF FORCES

In physics and mechanics, forces are represented by vectors. Simon Stevin, a Flemish mathematician and physicist, was the first one who explained the vector analysis.
A vector is a straight arrow pointing in the direction in which it acts. A vector is defined by:

- Magnitude
- Direction
- Point of Application (position)

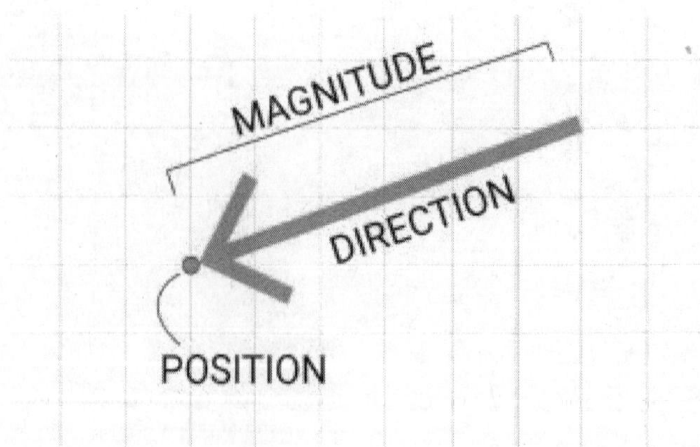

Figure 2-1: A vector is defined by magnitude, direction and point of application (position)

Forces may act on structural elements at one or multiple points. For example, a column that sits directly on a concrete slab applies a point load on a surface. Also, forces may be distributed uniformly (like the self-weight of a beam) or non-uniformly (like fluid pressure or snowdrift) along the length or surface of a structural element.

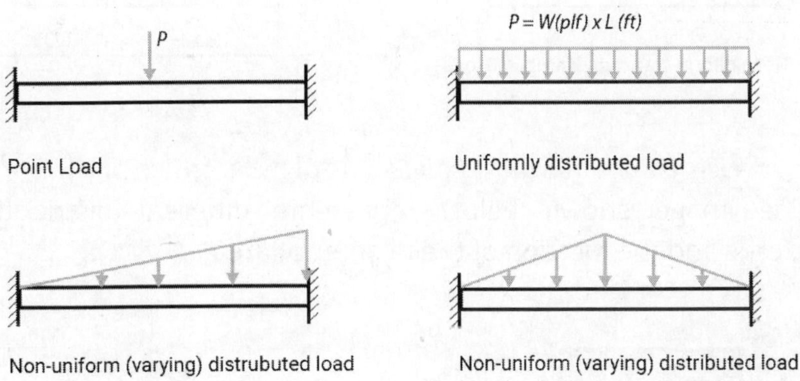

Figure 2-2: Distribution of load on a simple beam

The total magnitude of a distributed load can be represented as a single (resultant) load applied at the center of gravity. In two dimensions, the center of gravity is the same as the center of area of the distribution pattern.

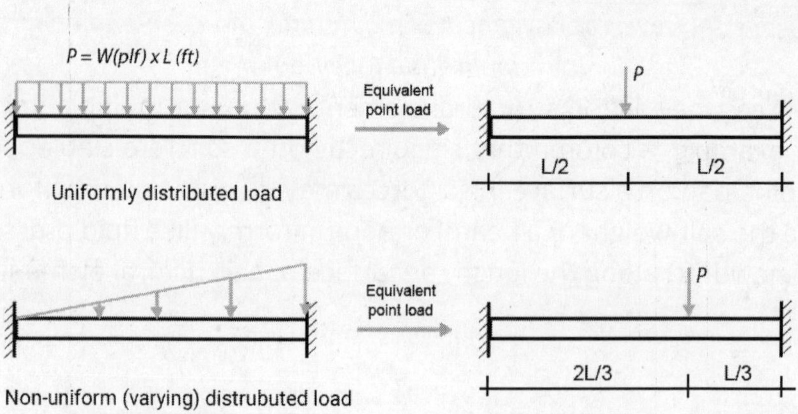

Figure 2-3: Equivalent point loads of two distributed loads

The image shown below represents different distribution patterns and the location of their center of area.

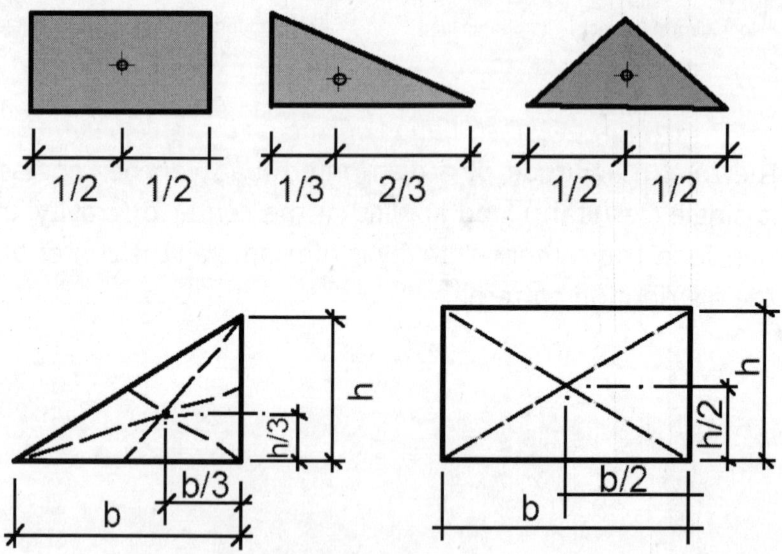

Figure 2-4: The center of gravity of different distribution patterns of loads

CHARACTERISTICS OF FORCES

Forces may cause:

- Tension: pulling apart.
- Compression: pushing together.
- Bending: applying equal and opposite couples in its own plane to an element; Or applying transverse forces to an element at some distance from its support.
- Torsion: twisting action from applying equal and opposite couples to the ends of a similar element in planes at a right angle to its axis.
- Shear: an action that tends to cause slipping of one part of an element on another.

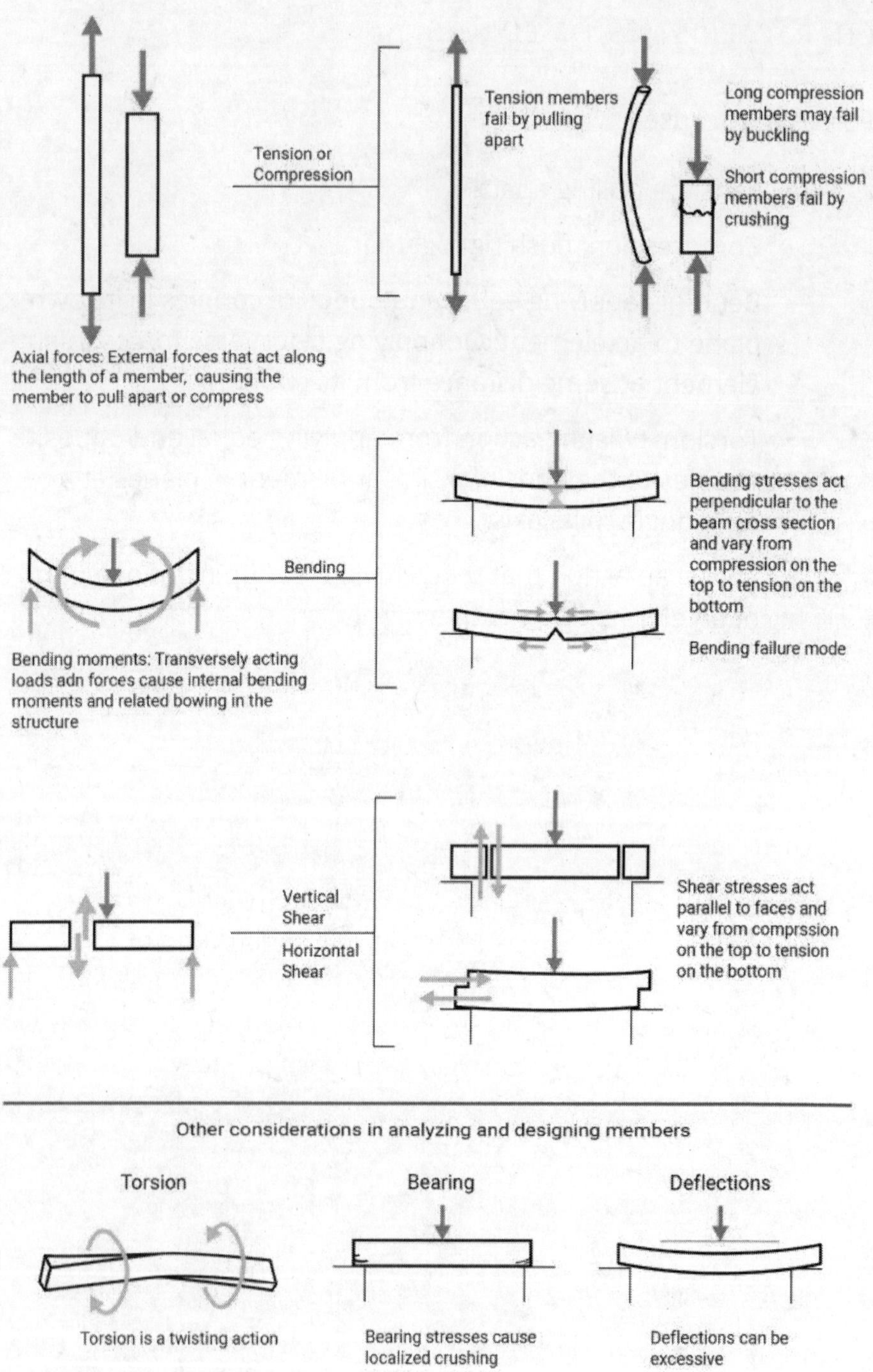

Figure 2-5: Different characteristics of loads.

Note: The "Buckling" effect in long compression members will be discussed in Chapter 11.

FORCE SYSTEMS

Force vectors may pass through a single point (concurrent) or be parallel (non-concurrent). Furthermore, force vectors may lay in a single plane (coplanar) or cannot be in a single plane (non-coplanar). Therefore, four different force systems can be identified:

- Concurrent – Coplanar
- Non-concurrent – Coplanar
- Concurrent – Non-coplanar
- Non-concurrent – Non-coplanar

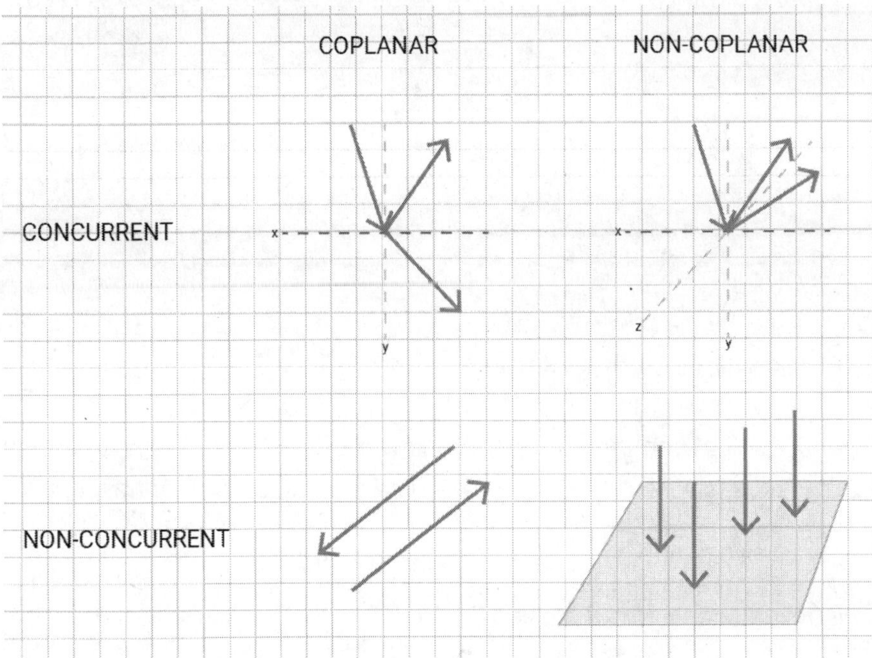

Figure 2-6: Different force systems

FORCE ADDITION

1. Adding inline forces

If force vectors are all inline, you can simply add their magnitudes to find the resultant. You can see an example of linear addition in a tug-of-war rope contest.

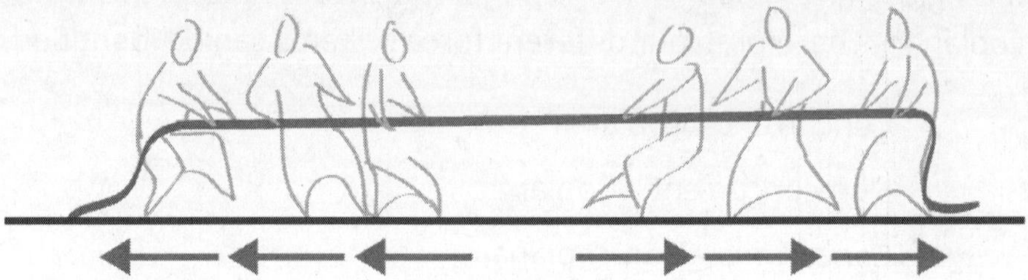

Figure 2-7: When the two force vectors are applied in two different directions and have the same magnitude the resultant equals to zero.

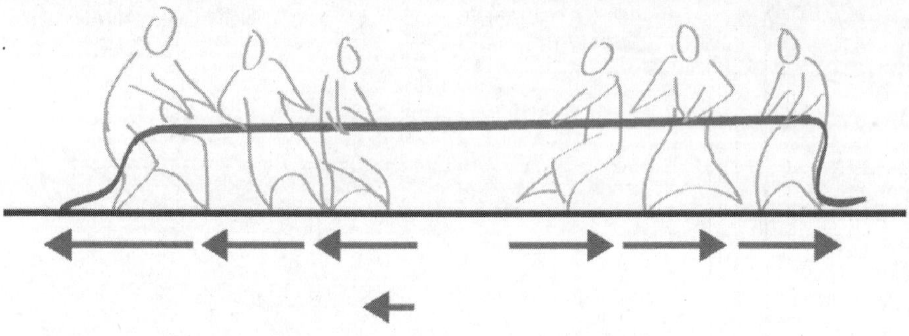

Figure 2-8: When the total of one set of force vectors are greater than the other set the resultant will be applied to the direction of the set with greater magnitude.

2. Adding orthogonal forces

Two orthogonal forces can be added together, simply by using the Pythagorean Theorem. The resultant equals to the hypotenuse.

You may refer to "Intro to the Pythagorean theorem" at Khan Academy to review the Pythagorean equation.

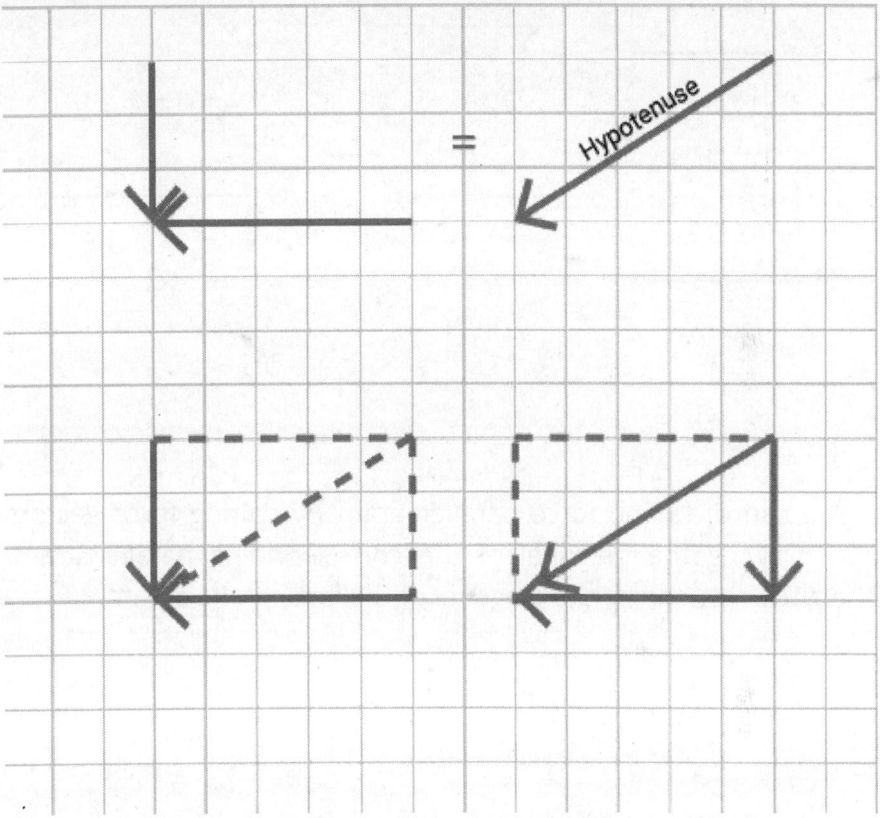

Figure 2-9: The resultant of two orthogonal forces equals the hypotenuse of the corresponding triangle

3. Adding two or more forces using graphic methods

You may add two or multiple force vectors by drawing either the **force polygon** or **force parallelogram** to find the resultant. In both cases, first, you should draw the force vectors to scale. Then, if you draw the force polygon, add the forces **head to tail**, and the resultant is the vector that closes the polygon **tail to head**.

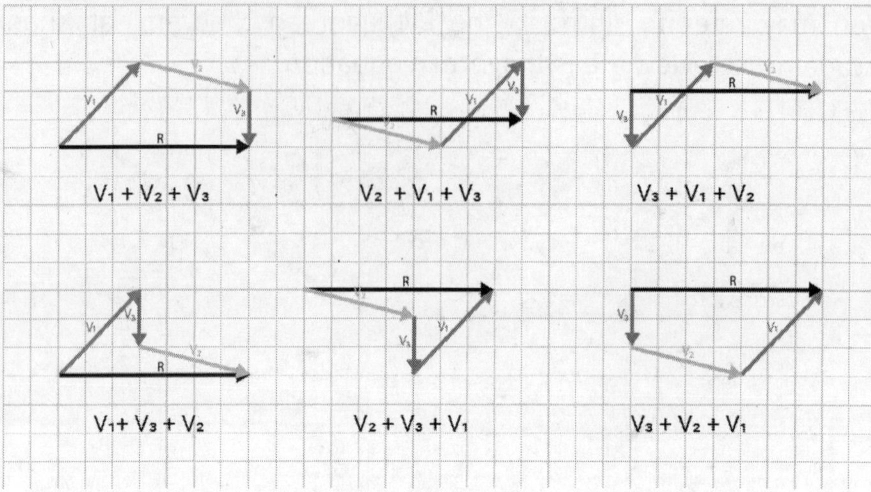

Figure 2-10: Adding two or more forces using the method of force polygon

You can draw the force parallelogram by pairing force vectors and drawing the diagonal of the corresponding parallelogram. Then, the resultant is the diagonal of the last parallelogram.

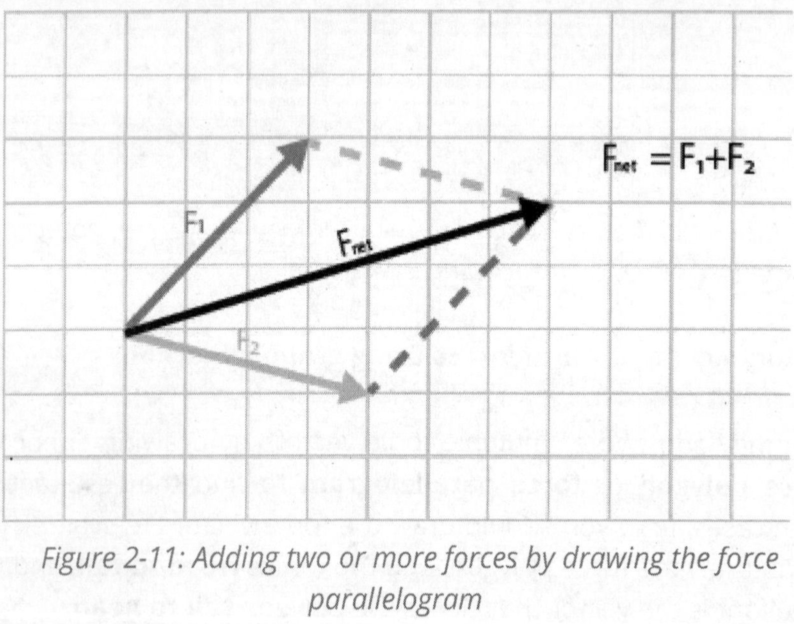

Figure 2-11: Adding two or more forces by drawing the force parallelogram

4. Adding two or more forces using an analytical method

When there are multiple force vectors, you may break each force vector into its horizontal and vertical components, sum all horizontal vectors together, sum all vertical components, and find the resultant of the orthogonal vectors. For example, F1, F2, and F3, shown below, are broken into their orthogonal components. Then, all horizontal components are added together, and all the vertical components are added together. Finally, the resultant of the two horizontal and vertical vectors can be calculated by using the Pythagorean theorem.

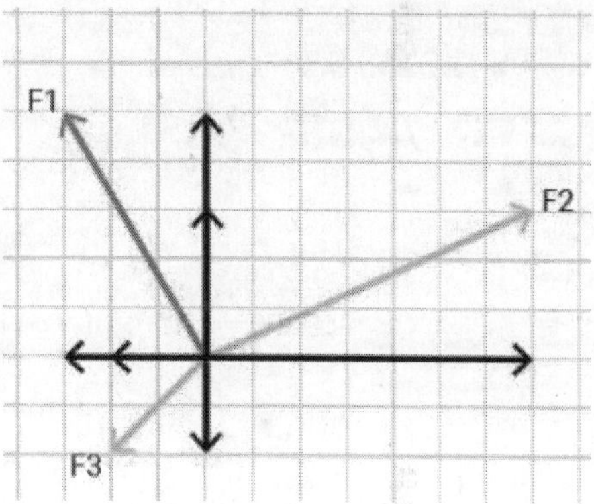

$\Sigma F_x = F1_x + F_{2x} + F_{3x}$,

$\Sigma F_x = -3 + 7 - 2 = 2$

$\Sigma F_y = F_{1y} + F_{2y} + F_{3y}$,

$\Sigma F_y = 5 + 3 - 2 = 6$

$F_{total} = \sqrt{(2^2 + 6^2)} = \sqrt{40} = 6.32$

If you would like to review the fundamentals of vectors analysis, see Khan Academy's module on Vectors.

The basics of orthogonal triangle trigonometry are represented below. Moreover, you may review right triangles & trigonometry at Khan Academy.

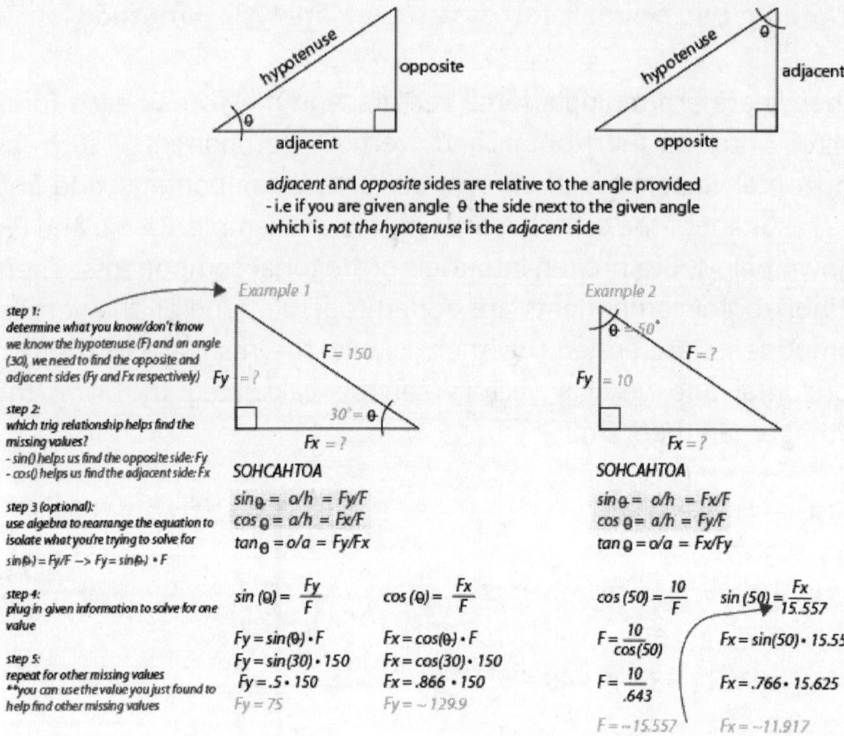

Figure 2-12: Basic review on solving orthogonal right-angle triangles

EQUILIBRANT

The equilibrant is a vector that has the same size as the resultant but is in exactly the opposite direction. Equilibrant balances the forces and makes them not push.

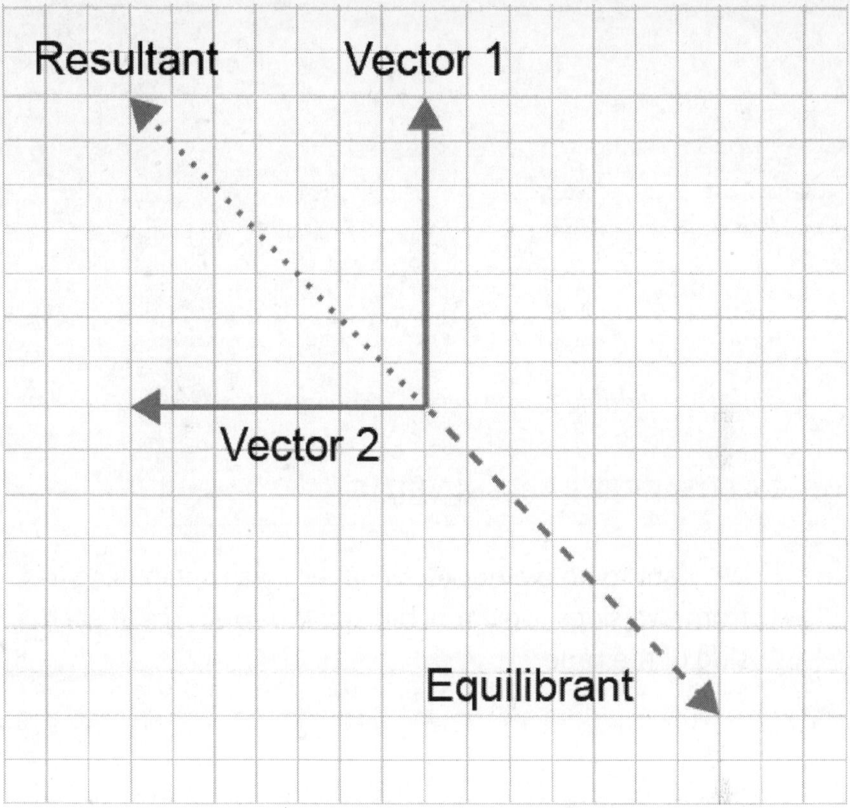

Figure 2-13: Resultant vs. equilibrant

Watch the following video in "Equilibrium of a Point" (https://www.youtube.com/watch?v=a2IX52UlCWE) to learn what condition must a point in a structure satisfy to remain in equilibrium.

One or more interactive elements has been excluded from this version of the text.

You can view them online here: https://pdx.pressbooks.pub/archistructures/?p=22#oembed-1

Video 2-1: Adding force vectors and equilibrium of a point

TOPICS FOR CRITICAL THINKING

- We want to hold a bucket of 50 lbs by a rope pulling it from two sides. How much is the tension force in each side of the rope?

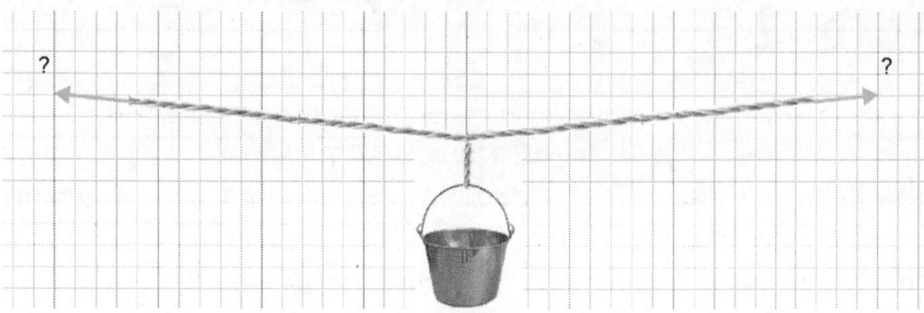

- Do you think that you can flatten a clothesline

(Image source: Clothes Line, https://www.flickr.com photos/ bmitd67/5045044902)

- Play around vector addition using PhET interactive simulator provided by the University of Colorado Boulder.
- Practice application of principles of vector addition using the Vector Guessing Game Interactive.
- What is the difference between deflection and bending

CHAPTER 3

Equilibrium

MOMENT OF FORCE

Sometimes the reaction to a force is not in the same line with the force and causes twisting. This twisting effect is called the moment of the force. Moment of the force depends on two parameters:

- force
- its distance between the force path and the point about which the moment is taken.

Thus, the moment of the force is defined as:

force × perpendicular distance between the force path and the point about which the moment is taken

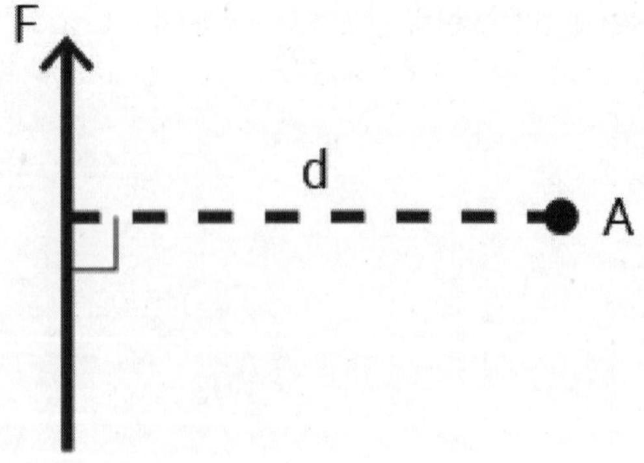

Figure 3-1: Moment of the force = force × perpendicular distance between the force path and the point about which the moment is taken

There is a sign convention for rotational forces. This sign convention does not relate to the direction of the force arrow but the direction of the rotation.

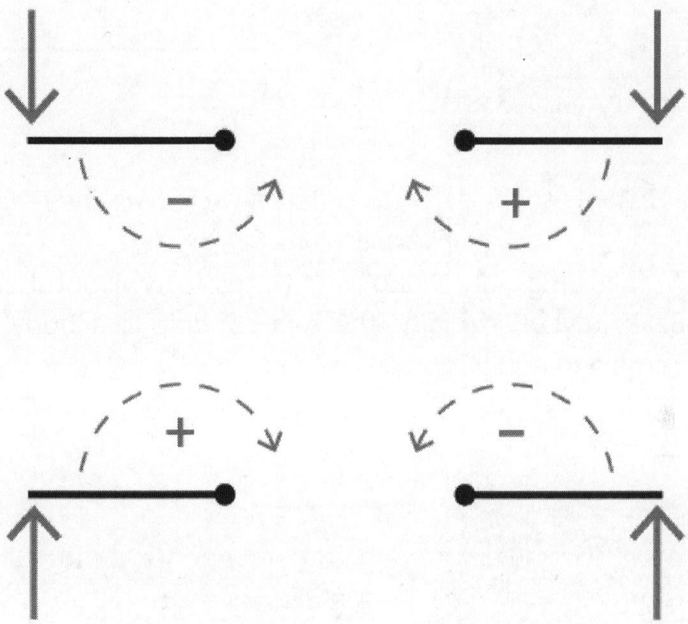

Figure 3-2: Sign convention for moment of the force

NEWTON'S FIRST LAW

An object at rest will remain at rest unless acted upon by an outside external net force.

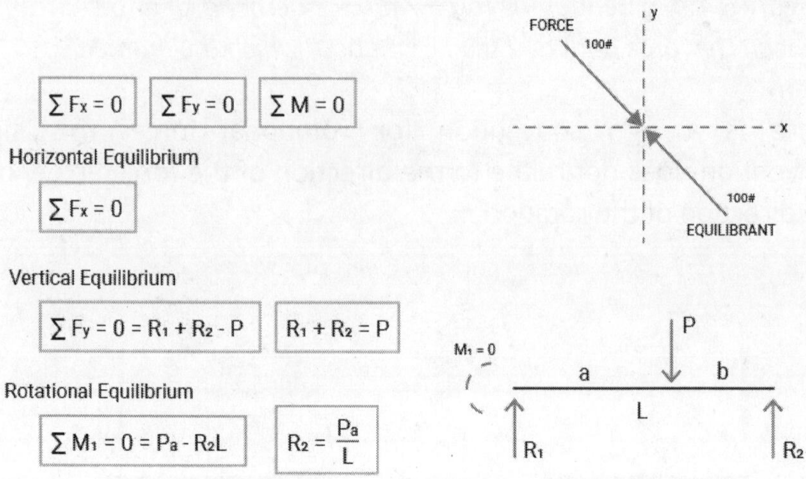

Figure 3-3: Application of Newton's first law in finding end reactions of simple beams

The following video (https://www.youtube.com/watch?v=6zXkYjmvLuI) explains the two conditions a body must satisfy to remain in equilibrium.

One or more interactive elements has been excluded from this version of the text. You can view them online here: https://pdx.pressbooks.pub/archistructures/?p=24#oembed-1

Video 3-1: Equilibrium of a body

You can see the rotational equilibrium in action in a balanced seesaw (lever) at your local playground. In the following image, the two individuals can determine the suitable length of the moment arm by adjusting their distances from the fulcrum to balance the seesaw.

Figure 3-4: Equilibrium in a balanced seesaw

Rotational equilibrium requires: $M_1 + M_2 = 0$

$M_1 = F_1 \times d_1$ $\quad M_1 = 25 \times 4 = -100$

$M_2 = F_2 \times d_2$ $\quad M_2 = 100 \times 1 = +100$

Stability against overturning

Watch the following video (https://www.youtube.com/watch?v=iDzp6xEAT2I) to learn how retaining walls work and how they resist sliding and overturning. This is similar to how the building as a whole needs to resist the hydrostatic and wind loads mentioned in Chapter 1.

One or more interactive elements has been excluded from this version of the text.

You can view them online here: https://pdx.pressbooks.pub/archistructures/?p=24#oembed-2

Video 3-2: Stability of retaining walls

Resisting gravity must be greater than the overturning moment: $M_R > M_O$

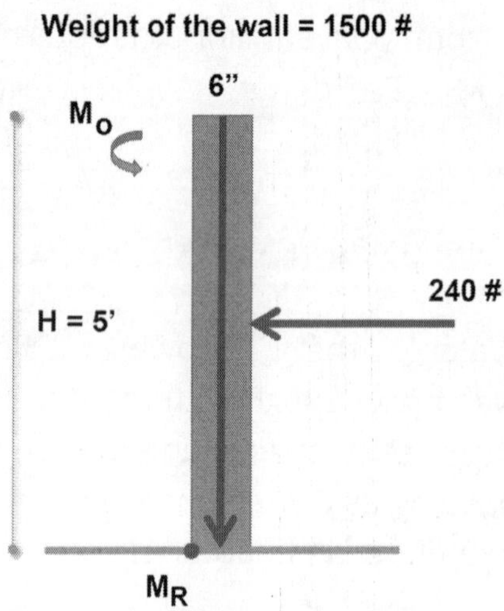

Example:

A concrete wall with a density of 150 pcf is expected to retain a mass of soil on one side. The distributed lateral load affecting the wall is equivalent to a point load of the soil at 24 psf. Study the wall stability against overturning and see if the wall stands the lateral load.

From the weight of the concrete wall:

M_R = F x d = 1500 × (3/12) = 375 lbs-ft

From the soil:

M_O = F x d = 240 × 2.5′ = – 600

Thus, the wall falls over!

ARCHIMEDES' PRINCIPLE OF THE LEVER

Archimedes' principle of the level indicates that two forces will balance at distances reciprocally proportional to their magnitudes.

$$F_1 \cdot a = F_2 \cdot b$$

$$F_1 = F_2 \cdot \frac{b}{a}$$

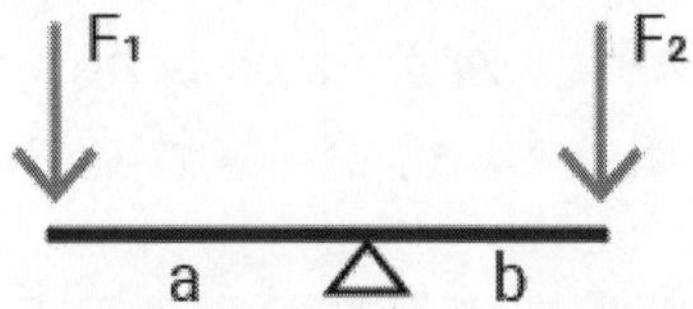

Figure 3-5: Archimedes' principle of the level

"Give me a place to stand, and I shall move the Earth." To learn more about Archimedes' idea and the amazing implications and uses of the lever, watch the following video (https://www.youtube.com/watch?v=YlYEi0PgG1g):

One or more interactive elements has been excluded from this version of the text. You can view them online here: https://pdx.pressbooks.pub/archistructures/?p=24#oembed-3

Video 3-3: How levers work

Archimedes principle is applicable in the determination of the end reactions of beams at their supports.

$$R_1 = P \cdot \frac{b}{L}$$

$$R_2 = P \cdot \frac{a}{L}$$

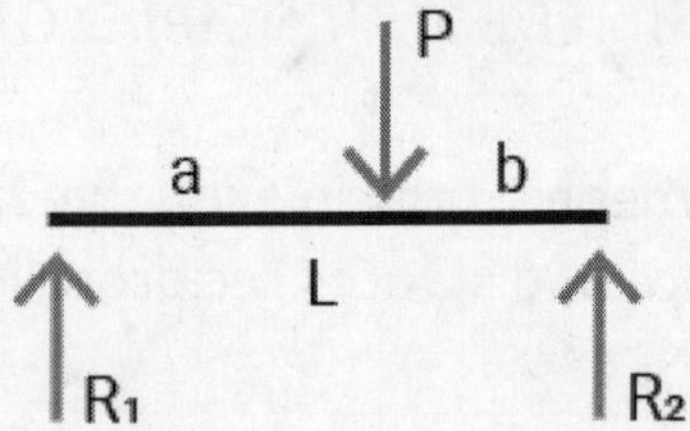

Figure 3-6: Application of Archimedes' principle of the level in finding end reactions in simple beams.

DETERMINATION OF THE END REACTIONS IN SIMPLE BEAMS

There can be three reaction vectors in simple beams upon the support conditions: horizontal, vertical, and rotational.

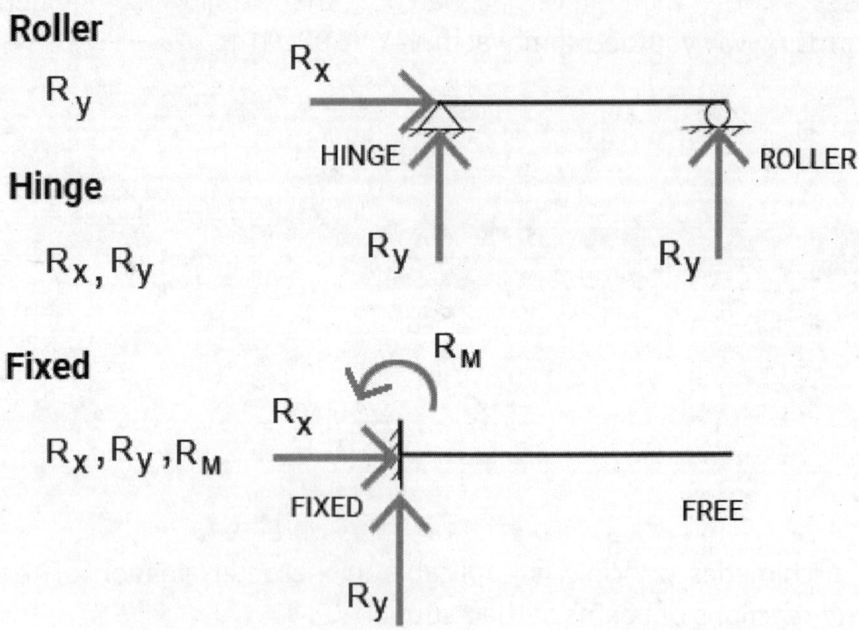

Figure 3-7: Reactions in beams with different support conditions

Figure 3-8: Hinged (pinned) connections in the structure of Renault Center in Britain

Figure 3-9: Hinged (pinned) connections in the structure of George Pompidou Center

Example 1

Using Newton's first law or Archimedes principle of the lever, we can study the following simple beam and find the shares of the two end supports of reacting to the point load P applied to the beam.

To proceed:

1. Label components of reactions. Consider the support condition and see if you should include vertical, horizontal, and rotational reactions.
2. Use the summation of moments about A to find $_B$.
3. Use the summation of moments about B to find $_A$ OR Find R_A load by summing vertical forces.

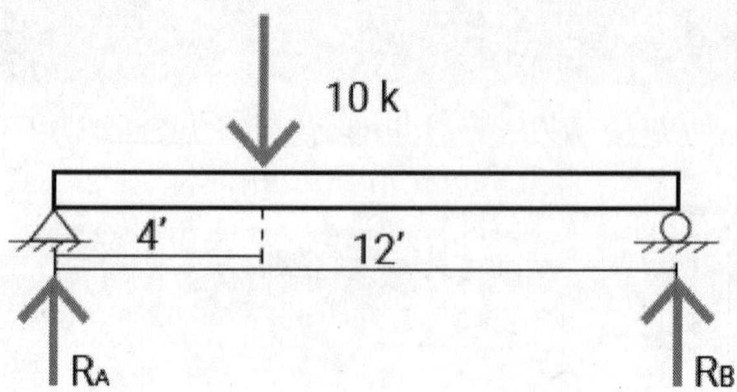

$\Sigma F_X = 0$ All horizontal forces must balance out to zero. No horizontal force!

$\Sigma F_Y = 0$ All vertical forces must balance out to zero.

$R_A + R_B - 10 = 0$

$R_A + R_B = 10$

$\Sigma M_1 = 0$ All rotational forces must balance out to zero.

$P \times a - R_B \times L + R_A \times 0 = 0$

$10 \times 4 - R_B \times 12 + 0 = 0$

$R_B = 40/12 = 3.33$ kip

$R_A = 10 - 3.33 = 6.67$ kip

Example 2

Find the end reactions at the supports of the simple beam shown below:

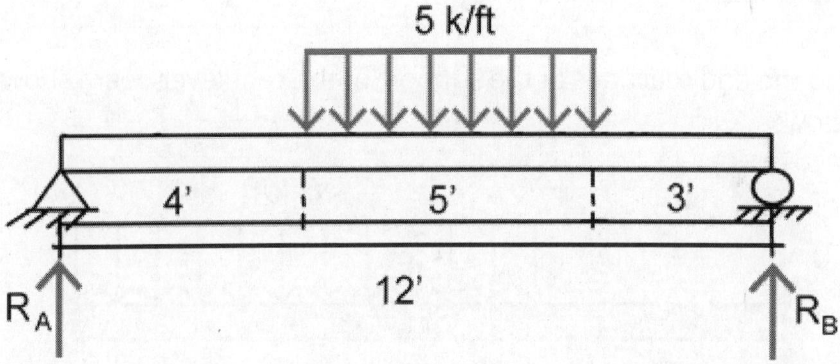

The effect of the distributed load applied to the beam can be studied by replacing it with its equivalent point load shown below:

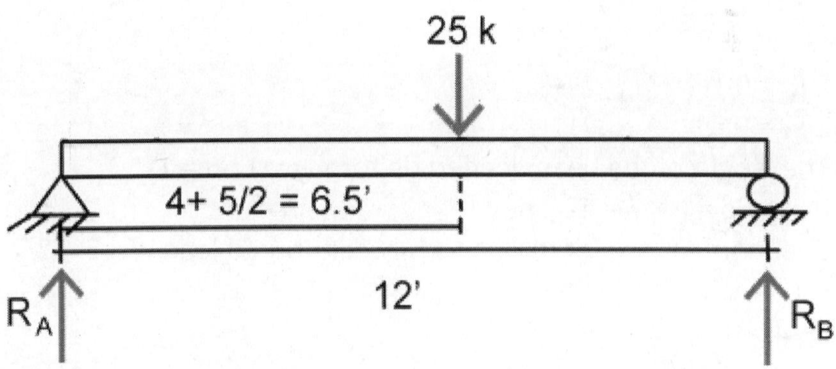

$\Sigma F_X = 0$ All horizontal forces must balance out to zero.

No horizontal force!

$\Sigma F_Y = 0$ All vertical forces must balance out to zero.

$P= w \times d = 5 \times 5 = 25$ k

$R_A+R_B -25 = 0$

$R_A + R_B = 25$
$\Sigma M\_A = 0$ All rotational forces must balance out to zero.
$25 \times 6.5 - R_B \times 12 + R_A \times 0 = 0$
$R_B = 162.5 / 12 = 13.54$ k
$R_A = 25 - 13.54 = 11.46$ k

Example 3

Find the end reactions at the support of the cantilever beam shown below:

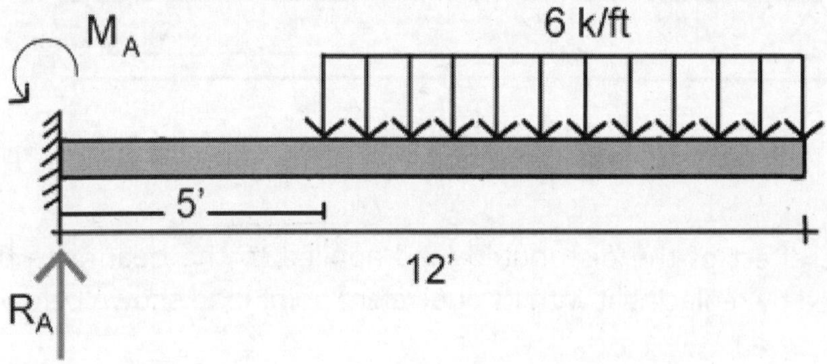

$\Sigma F_X = 0$
$\Sigma F_Y = 0$
The length of the beam loaded uniformly = 12-5 = 7
$P = w \times d = 6 \times 7 = 42$ kips
The moment arm of the uniformly distributed load w = 5 + 7/2 = 8.5
$R_A - 42 = 0$
$R_A = 42$
$\Sigma M_A = 0$ All rotational forces must balance out to zero.
$42 \times 8.5 - M_A = 0$
$M_A = 357$ kips. ft

TOPICS FOR CRITICAL THINKING

- Use a beam calculator interface to check your calculations when solving the reactions at the supports of a beam.

CHAPTER 4

Catenary Cables and Arches

CATENARY CABLES

GEOMETRY AND BASIC PRINCIPLES

Cables are structural elements that can hold a great amount of tensile force with a relatively small cross-sectional area. Thus, cables are considered highly efficien structural components, and one of the most economical ways to span a large distance is employing a cable system. However, cables can resist only axial tensile forces and not any compression or bending moment. Since cables have a minimal cross-sectional area, they are flexible elements and change their shapes under different load conditions to reach equilibrium. For example, a cable under a simple point load forms two straight lines meeting at the point of application of the load. Cables under a uniformly distributed load sag in a catenary shape or a parabola.

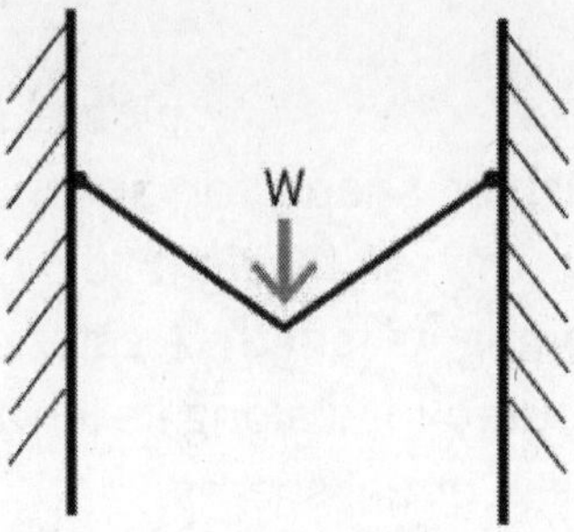

a) Simple concentrated load --- triangle

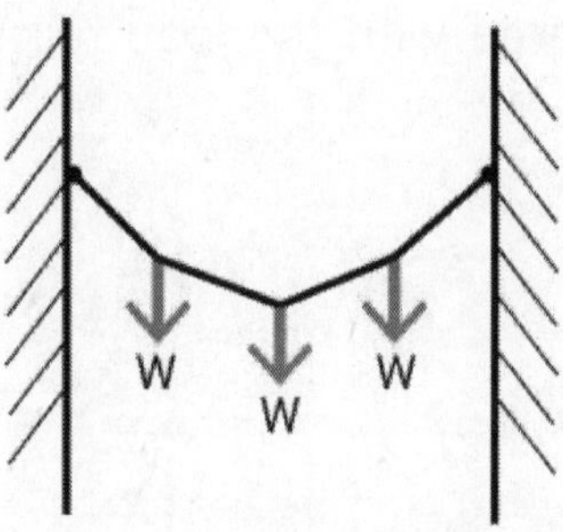

b) Several concentrated loads --- polygon

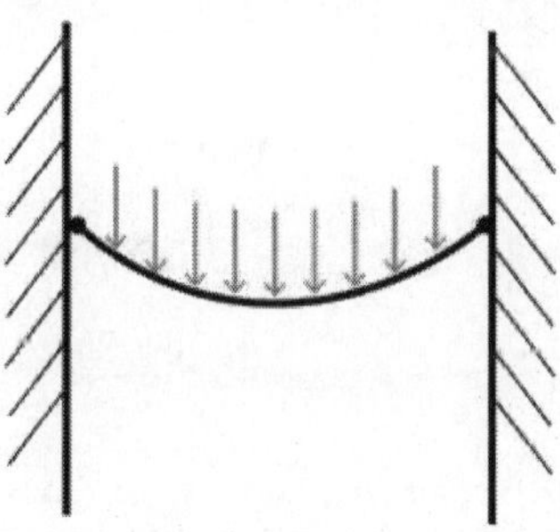

c) Uniform loads (along the cable length) — catenary

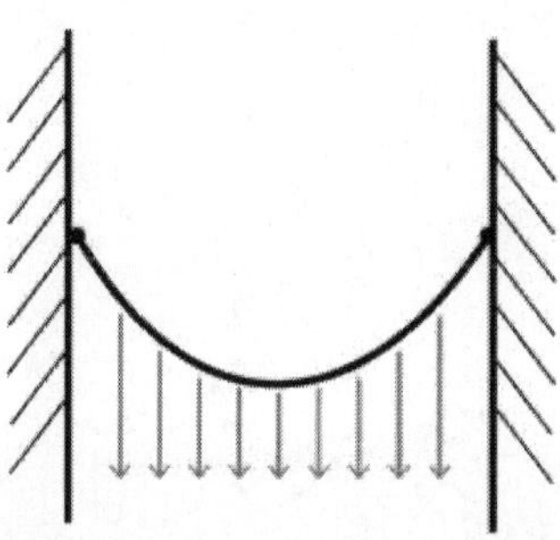

d) Uniform loads (horizontally) parabola

Figure 4-1: Forms of cables under different load conditions

A catenary is a funicular shape for an unloaded cable and is determined solely by the self-weight of the cable, which is uniformly distributed along its length. A catenary cable sags under such a uniformly distributed load along its length, and transfers the load to its two supports in equal shares.

In contrast, a parabola is a funicular shape of a suspension cable loaded uniformly across its span. Although, the self-weight should be insignificant compared with the load to give a pure parabola shape to a cable.

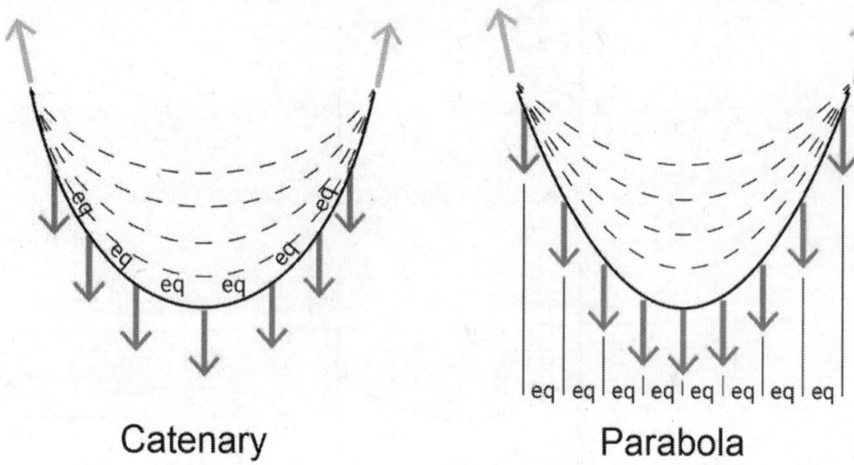

Figure 4-2: A catenary cable vs. a cable with a parabola shape

Where the sag-to-span ratio is greater than 5, the two shapes are nearly identical, and mathematically, it is simpler to utilize a parabola for analysis.

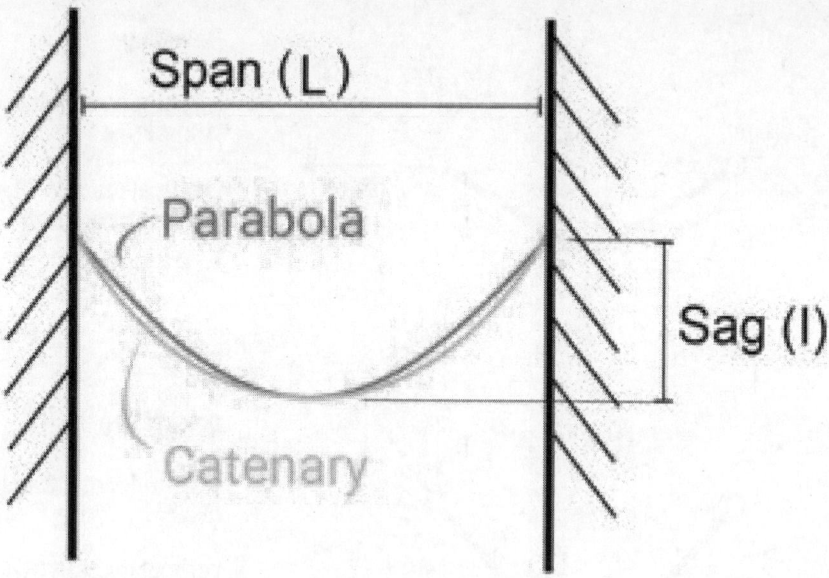

Figure 4-3: The difference of a catenary shape and a parabola

The load and the sag-to-span ratio determine the tensile force in a cable. Thus, in catenary cable structures, the sag-to-span ratio is a primary structural design consideration. The sag-to-span ratio influences:

- Cable forces: they are inversely proportional to sag.
- Inward thrust
- Cable length and the cable diameter: as the cable length decreases, a greater diameter is required.
- The column or tower (mast) height and the compressive forces

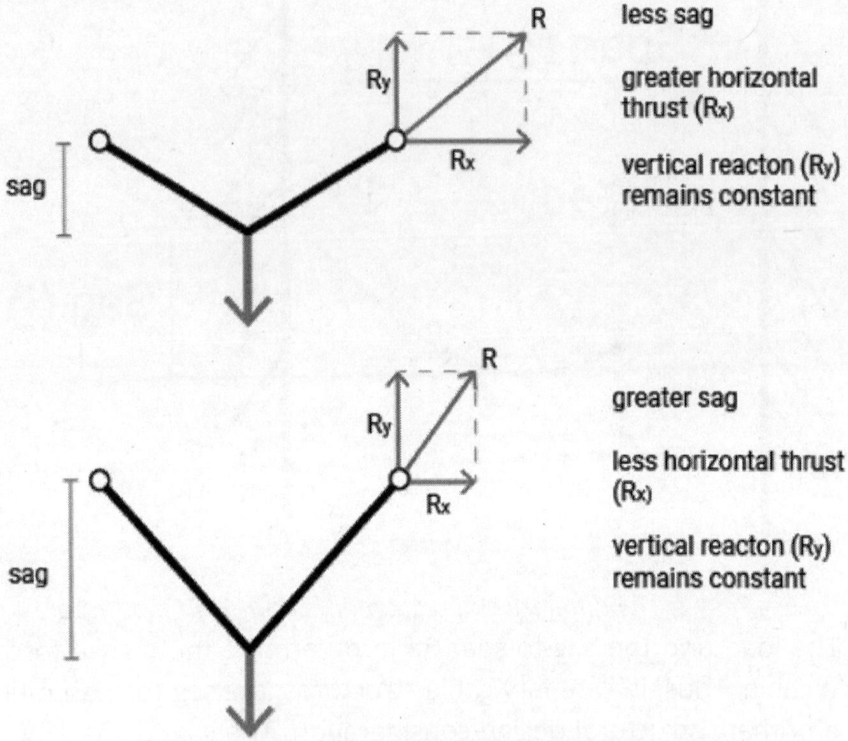

Figure 4-4: The relation between the horizontal thrust and the sag-to-span ratio in a loaded cable

For a uniformly loaded parabolic cable, the optimum sag-to-span ratio is 33%.

Most of the suspended roof structures (where cables are used for building the roof) have a sag-to-span-ratio of 1:8 to 1:10.

CLASSIFICATION OF THE TOPOLOGY OF CABLE STRUCTURES

Classification of cable structures allows a better understanding of their behavior and the determination of suitable design techniques. Cable structures can be classified into three main categories:

1. Single-curvature structures: A series of parallel catenary cables span between primary supports and hold a deck directly or hold a series of secondary vertical cables that carry the deck. A well-known example of this type of cable structure is the Dulles Airport Terminal. There are several suspension bridges where single-curvature cables shape the main configuration of the structural system. Akashi Kaikyo Bridge in Japan, Nansha Bridge in China, Verrazzano-Narrows Bridge in New York, Golden Gate Bridge in San Francisco, and Mackinac Bridge in Michigan are suspension bridges with single-curvature configurations
2. Double-cable structures: Some stabilizing cables are added below the primary suspensions cabless to resists the wind uplift. In Utica Memorial Auditorium, two layers of pre-tensioned cables span between an outer compression ring and a central tension ring.
3. Double-curvature structures: Some stabilizing cables are added below the primary suspensions cables to resists the wind uplift. Similar to a saddle shape, the primary suspension cables sag between the supports and cover the span. The stabilizing cables run in a perpendIcular direction with an opposite curvature. The roof of Dorton Arena in Raleigh is composed of a double-curvature cable structure and two compressive arches.

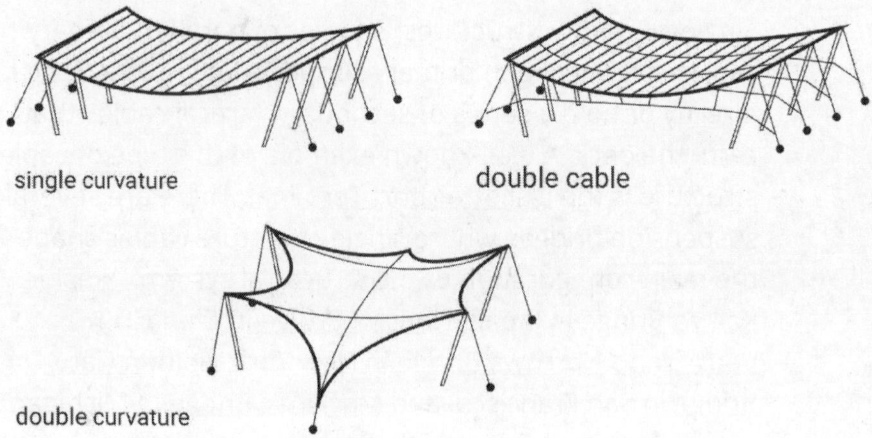

Figure 4-5: Different types of cable structures

Watch the following video (https://www.youtube.com/watch?time_continue=1&v=caTaBeKUh-U&feature=emb_logo) on suspension bridges and review what is discussed so far.

One or more interactive elements has been excluded from this version of the text. You can view them online here: https://pdx.pressbooks.pub/architructures/?p=26#oembed-3

Video 4-1: Structural behavior of suspension bridges

MATERIALITY

Cables can be of mild steel, high strength steel, stainless steel, polypropylene, fiberglass, and carbon. High-strength steel cables are the most reliable and economical material for linear tensile members. Structural cables are made of a series of small strands twisted or bound together to form a much larger cable. Usually, the steel used in cable structures has breaking stresses that exceeds 200,000 psi which is 4 times the strength of structural steel sections common in building construction.

THE PRINCIPAL ELEMENTS OF A CABLE-SUPPORTED STRUCTURE

A cable-supported structural system includes main cables, vertical supports or towers, anchorages, and stabilizers. Vertical supports may be masts, diagonal struts, walls, vertical or sloping piers. Vertical supports keep the cables above the ground and provide essential reactions.

Since the main cables are not placed vertically, the carried axial forces have a horizontal component. This horizontal thrust should be resisted by a suitable anchorage system. Therefore:

1. The main cables of a suspension bridge may be attached to the ground and be earth-anchored.

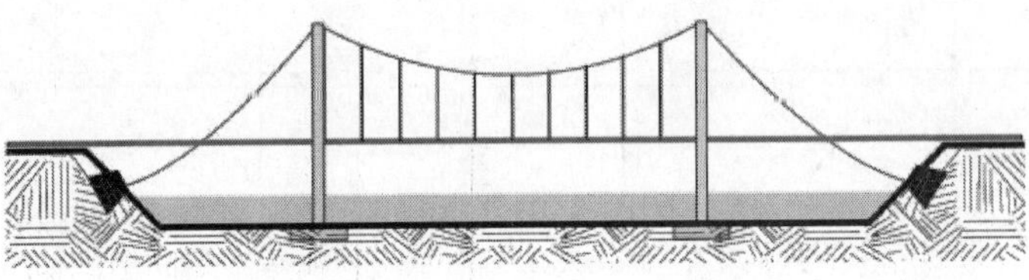

Figure 4-6: An earth-anchored suspension bridge

2. In a self-anchored suspension bridge, the main cables may be attached to the end of the road deck.

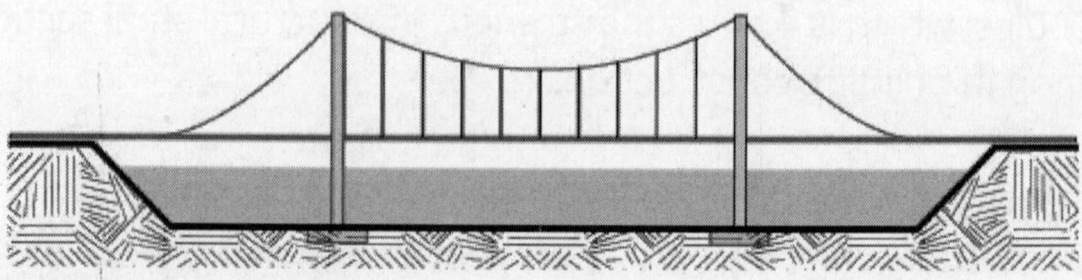

Figure 4-7: A self-anchored suspension bridge

Another type of self-anchoring bridges is where multiple primary cables hold the deck and their connection pattern may resemble a harp, a fan, or a star.

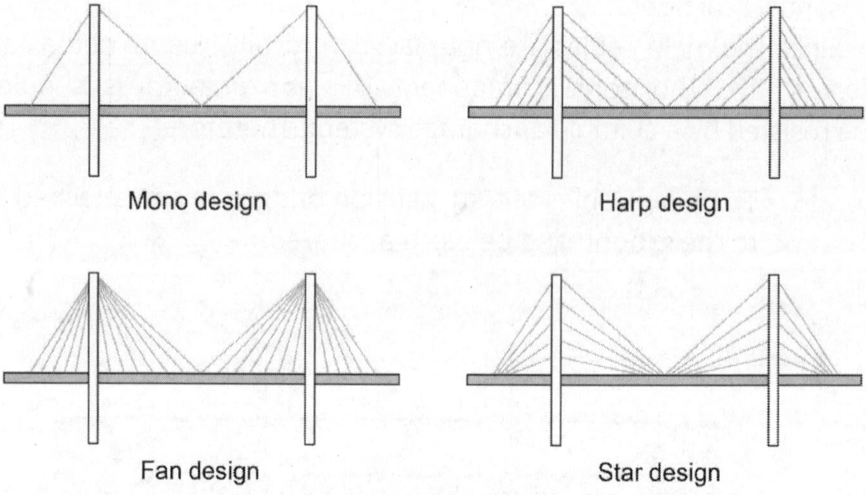

Figure 4-8: Different patterns of connecting cables to the bridge deck

Cables must change shape whenever the loads shift in location or amount. This flexibility brings practical difficulties Thus, we must stiffen the cables by means of a beam, truss and stabilizers. Uneven loading such as wind, and vibration may be destructive. Considering a clothes-line you may propose different ideas for stabilization of the cable against wind loads.

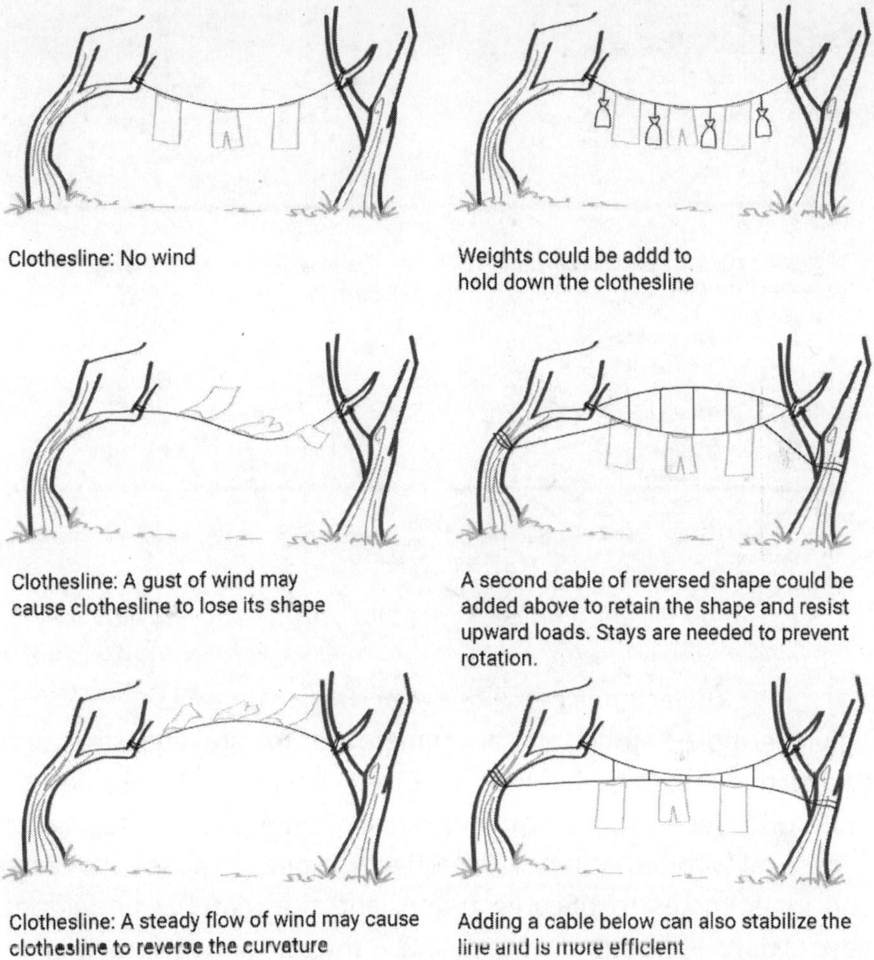

Figure 4-9: Different ideas for stabilizing a cable against wind loads

Usually, designers use the dead weight of the roof/deck construction, a rigid surface that includes the main cables, a set of pre-tension cables with reverse curvature from the main cable, or restraining cables to stabilize cable-stayed structures. In the Dulles Airport Terminal, you can see some of these thoughtful ideas are applied to stabilize the cable-stayed roof structure.

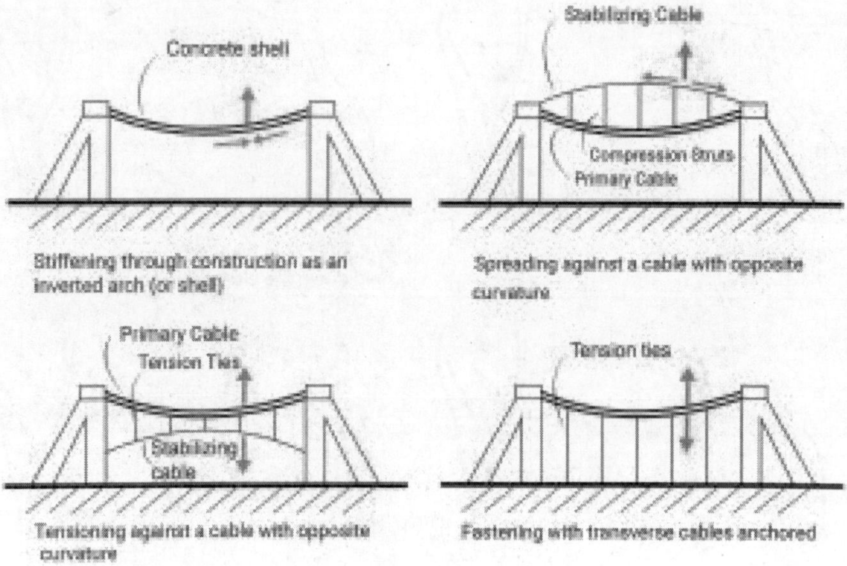

Figure 4-10: Different ideas for stabilizing a cable-stayed roof structure. Note: Primary cables carry gravity loads and secondary cables stabilize the bridge against uplift forces.

Furthermore, stabilizers are employed to prevent structural destruction due to probable resonance. All materials have a natural molecular vibration or frequency range. If an outside force acts on a material within that natural frequency range, it causes the outer and inner forces to become tuned, and the material undergoes destruction. Watch video 4-2 (https://www.youtube.com/watch?v=n9ULMIjvSIg) to learn more about the effect of resonance on building structures.

One or more interactive elements has been excluded from this version of the text.

You can view them online here: https://pdx.pressbooks.pub/archistructures/?p=26#oembed-4

Video 4-2: The effect of resonance on structures

Tacoma Narrows Bridge was a suspension bridge in the state of Washington that collapsed on November 7, 1940. The bridge collapsed because normal speed winds produced aeroelastic flutter that matched the bridge's natural frequency. The collapse of the bridge was recorded on film by Barney Elliott, owner of a local camera shop. See the collapse of the Tacoma Bridge in the following video (https://www.youtube.com/watch?v=XggxeuFDaDU&t=1s).

One or more interactive elements has been excluded from this version of the text. You can view them online here: https://pdx.pressbooks.pub/archistructures/?p=26#oembed-1

Video 4-3: Tacoma Bridge destruction

In addition to analyzing the principal elements, a cable-supported structure should be finely studied regarding corrosion protection, in-service inspection, fire resistance, and effects on the structure due to the removal and replacement of cables during the lifetime of the structure.

CATENARY ARCHES

Arches have been used in many subtle historical constructions. A catenary arch redirects forces into axial compression to span an opening. How do arches work? Watch the following video (https://www.youtube.com/watch?time_continue=3&v=JlL6ZHChhQE&feature=emb_logo) to learn about their behavior and discover an intriguing relationship between arches and hanging chains.

One or more interactive elements has been excluded from this version of the text. You can view them online here: https://pdx.pressbooks.pub/archistructures/?p=26#oembed-2

Video 4-4: Structural behavior of arches

As you learn from the video, a funicular arch is the inverted compressive equivalent of a suspension cable. In the 17th century, Robert Hooke studied the catenary forms of a hanging chain and discovered this principle. A catenary arch is subject to carry only axial compression and not bending forces.

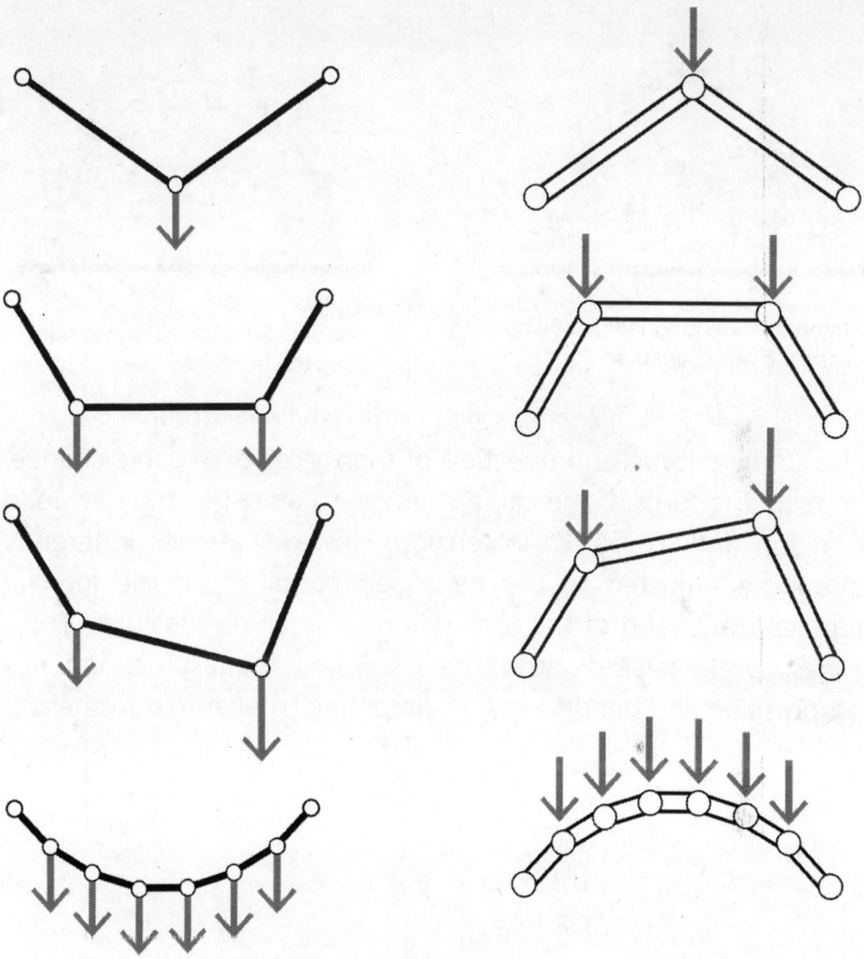

Figure 4-11: The relation between a funicular arch and its corresponding funicular cable

Suppose the applied load is distributed uniformly along the length of an arch. In that case, the funicular arch should be equivalent to an inverted catenary cable to carry the loads only in axial compression. On the other hand, if the load is distributed uniformly along the span, the form of the funicular arch should be equivalent to a parabola.

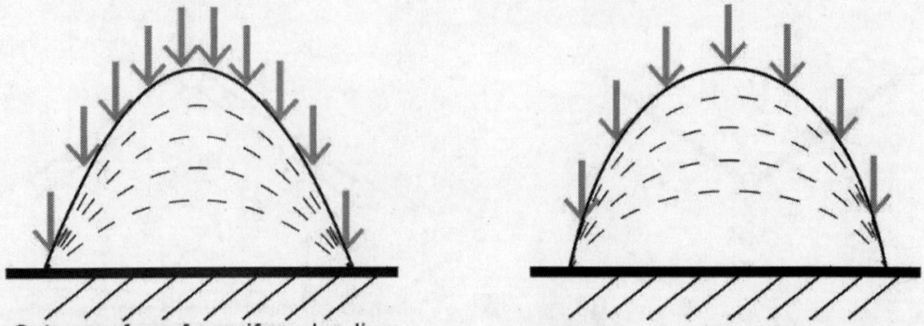

Figure 4-12: A catenary arch vs. an arch with a parabola shape

If the magnitude and direction of the forces on a cable change, the resulting form of the cable changes to adjust to the new load condition and to remain in tension. However, arches inherently cannot be adjusted to the new load condition. If the loading changes the shape of the arch will no longer be funicular. Thus, arches and related structural systems usually experience compression and bending at the same time (similar to thin shells).

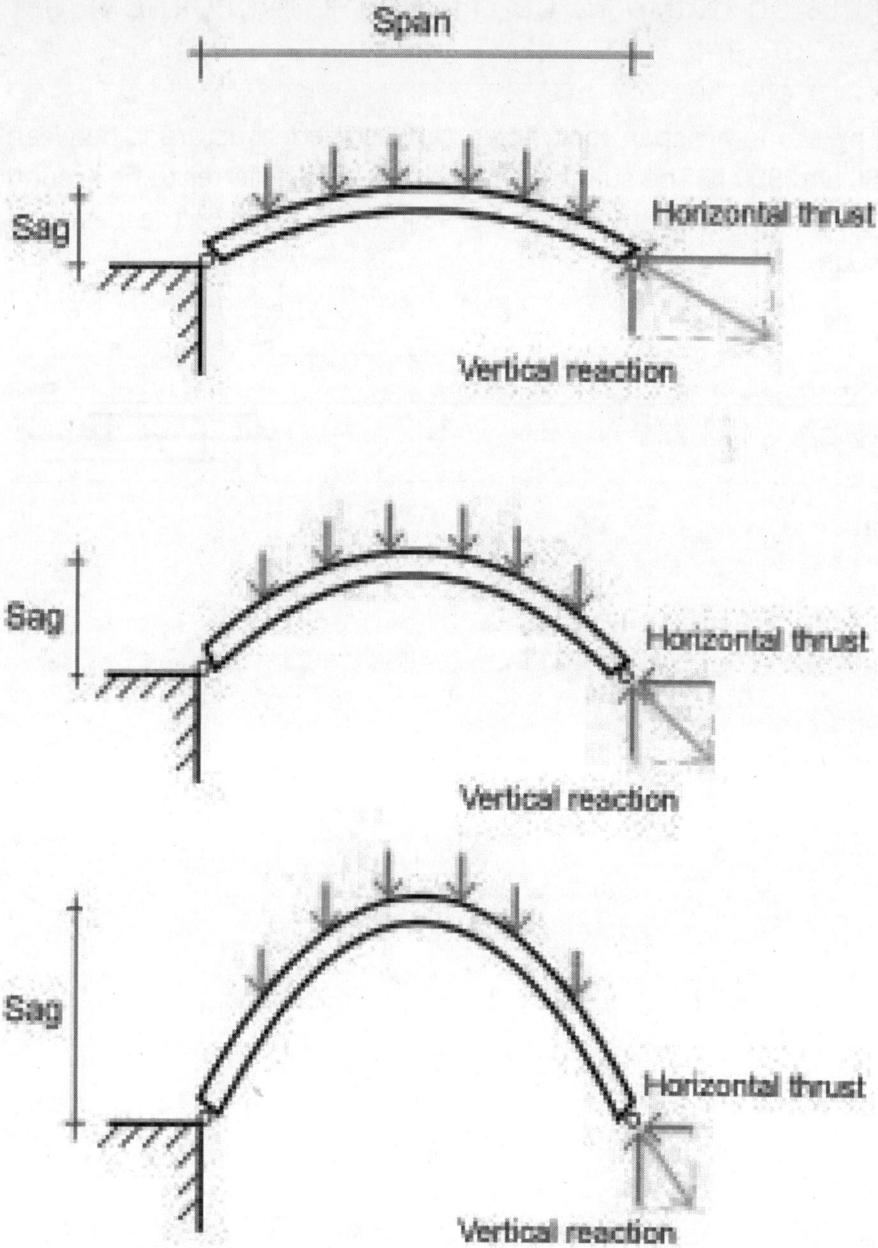

Figure 4-13: The relation between the horizontal thrust at the supports of an arch and its rise-to-span ratio

RULE OF THUMB TO DETERMINE THE PROPORTIONS OF STRUCTURAL COMPONENTS

The economic span for using a cable-stayed structure is between 50 and 500 ft. The suitable sag to span ratio is better to be around 1:10. The cable diameter can be 1:8 into 3 ft based on the loads and span.

Table 4-1: The economic span of a cable-stayed structure [2]

STRUCTURAL SYSTEM		Pages	Span Range 10' 3m	20' 6m	30' 9m	50' 15m	100' 30m	200' 60m	300' 90m	500' 150m
STEEL	Arches and Vaults									
	Cable-Stayed									
	Suspension									

CHAPTER 5

Trusses

GEOMETRY AND BASIC PRINCIPLES

A truss represents a structural system whose elements are two-force members arranged in a planar triangular pattern and each member is either in tension or compression. The stability of a truss relates to its triangular shape. By definition, trusses have pinned joints and concurrent straight members and have to be loaded through their joints. This means that a frame structure with rigid joints is not considered a true truss. A Vierendeel frame is a well-known example of such a frame structure with rigid joints that cannot be considered as a true truss. In reallty, a combination of bolting and welding is used to make the joints. Thus, joints become somewhat rigid connections that develop some moment resistance. Bending stresses, however, are often relatively small in comparison to those resulting from tension and compression.

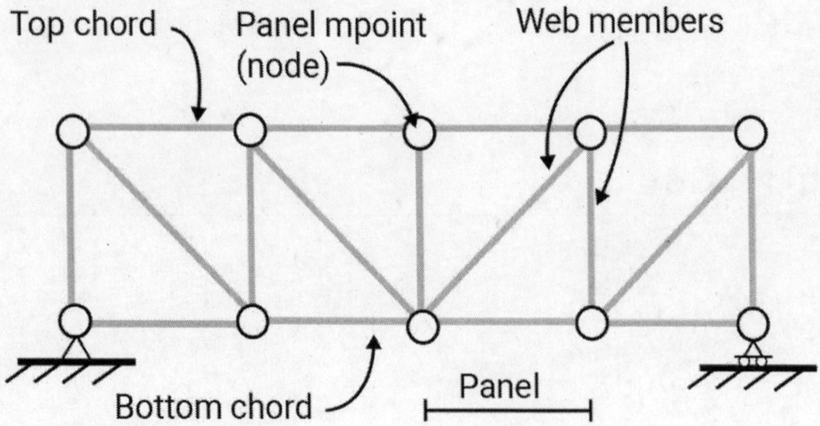

Figure 5-1: Truss components' nomenclature

A center-loaded cable with pinned roller supports cannot resist the horizontal thrust and is unstable. By adding a compressive strut, the system resists the thrust internally to form a simple truss.

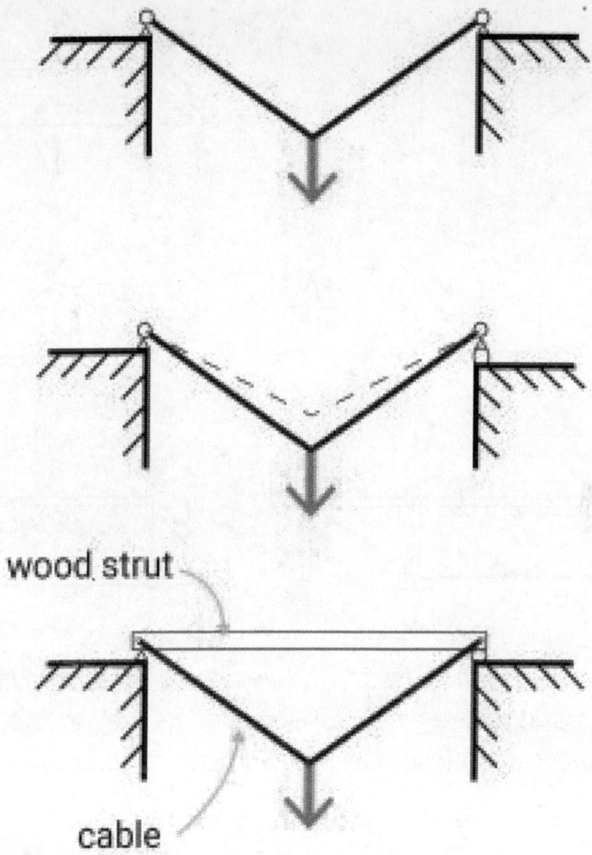

Figure 5-2: The process of creating a simple truss module

This simple truss module is used to form complex forms of trusses where compressive elements are made of rigid struts (solid lines), and tensile strengths are substituted by cables (dash lines).

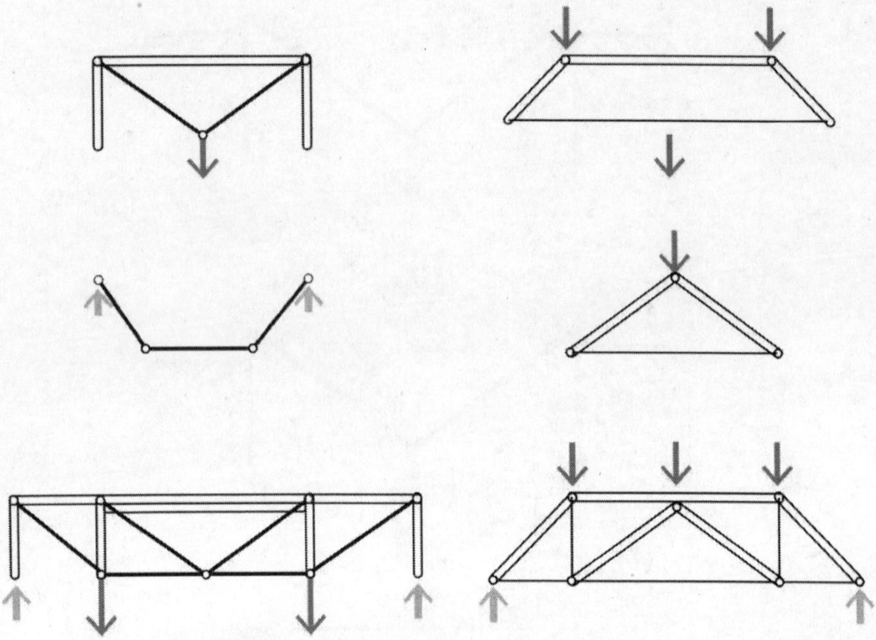

Figure 5-3: A simple truss module can be used to form complex forms of trusses where compressive elements are made of rigid struts (double lines), and tensile strengths are substituted by cables (single lines).

STABILITY AND DETERMINACY IN TRUSSES

The first step in designing a truss is the analysis of its stability, its internal and external determinacy or indeterminacy. Stability in trusses refers to their ability to maintain their configuration while resisting loads applied to their joints. For a stable truss the equilibrium conditions ($\sum F_x = 0$, $\sum F_y = 0$, $\sum M = 0$) are always satisfied regardless of the position or direction of the applied loads. In other words, if we can find even a single loading case for which the equilibrium equations cannot be satisfied, then we must conclude that the truss is unstable.

A stable truss may be either statically determinate or indeterminate. When a member is added to a stable truss or the number of support reactions is greater than the number of equilibrium equations, the truss is still stable. But if the number of unknown forces is greater than the number of equilibrium equations, the truss is considered statically indeterminate. There is a nifty method for the identification of stability and determinacy in trusses.

For:

- j, joints
- m, members
- r, reactions (restraints)
- k, **k= 2j-r**

Then,

If $m < k$, the truss is unstable

If $m = k$, the truss is stable and determinate

If $m > k$, the truss is stable and indeterminate

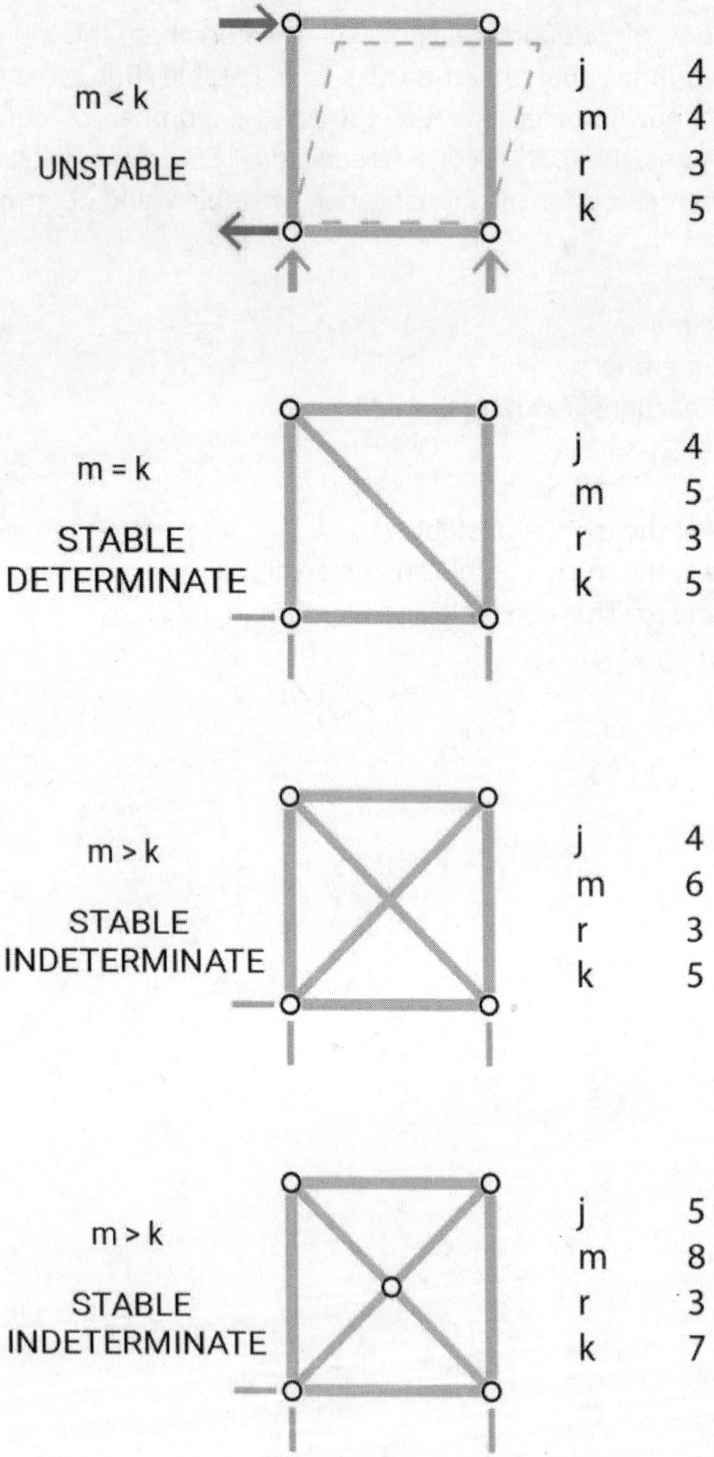
m < k
UNSTABLE
j 4
m 4
r 3
k 5
m = k
STABLE
DETERMINATE
j 4
m 5
r 3
k 5
m > k
STABLE
INDETERMINATE
j 4
m 6
r 3
k 5
m > k
STABLE
INDETERMINATE
j 5
m 8
r 3
k 7

Figure 5-4: Identification of stability and determinacy of trusses by using the k= 2j-r equation

This equation is helpful to identify the internal stability and determinacy of trusses. However, it is not sufficient to identif the external stability and determinacy. Visual inspection and intuitive sense must also be utilized to make the final assessment. For example, in the truss shown in figure 5-5 (b), m=k=13 and is expected to be stable and determinate. However, the truss is held by three roller supports and is horizontally unstable.

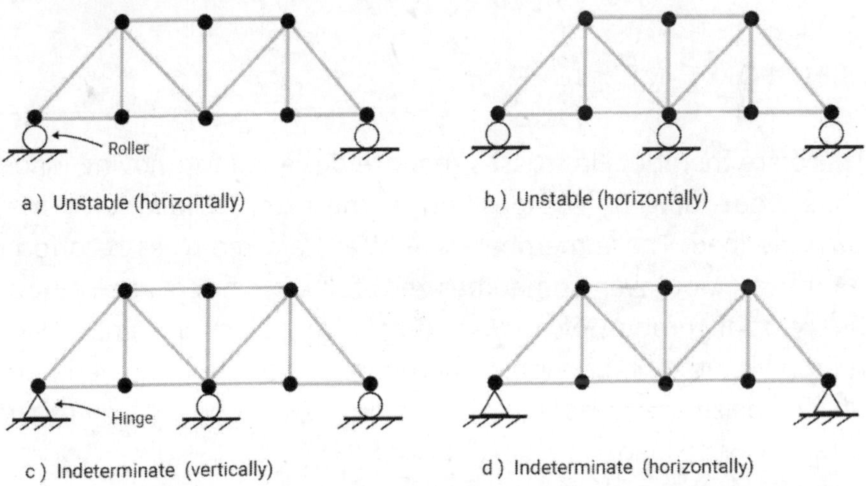

Figure 5-5: Examples of indeterminate or externally unstable trusses

Watch the following video (https://www.youtube.com/watch?v=w7rAiqzlanQ) to further investigate stability and determinacy in trusses.

One or more interactive elements has been excluded from this version of the text.

You can view them online here: https://pdx.pressbooks.pub/archistructures/?p=28#oembed-3

Video 5-1: Stability and determinacy in trusses

ZERO-FORCE MEMBERS

There are members in trusses that are designed for moving loads. Thus, depending on the position of the external load, they may carry no load. These members are often referred to as zero-force members. Moreover, sometimes zero-force members are included in the configuration of a truss to split long slender compression members into two or shorter members (brace the long members) and stabilize the truss against buckling. These bracing members often carry no load.

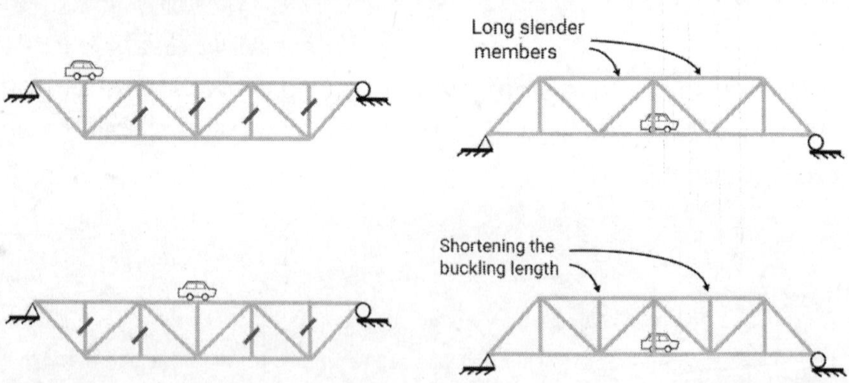

Figure 5-6: Zero force members in trusses under moving loads

For trusses to be in the state of equilibrium, the sum of the forces acting at each joint must be zero. By placing and rotating the coordinate system at a joint, we can study if there is only one force acting in either x or y-direction. That single force must be zero, and the member associated with it is a zero-force member. Watch the following video (https://www.youtube.com/watch?v=xSEScMN6bpo&t=89s) to practice this method of identification of zero-force members.

One or more interactive elements has been excluded from this version of the text. You can view them online here: https://pdx.pressbooks.pub/archistructures/?p=28#oembed-1

Video 5-2: How to identify zero force members in trusses

Truss analysis by the method of joints

Resolution of joints is one of the first methods for truss analysis. This method involves the following steps:

1. Solve reactions (all external forces)
2. Inspect for zero force members (Γ's & L's)
3. Cut a Free-body-diagram (FBD) of one joint. A free-body diagram of a joint consists a diagrammatic representation of the joint and all the forces acting on it.
4. Show forces as orthogonal components
5. Solve with ΣF_x and ΣF_y (no ΣM)
6. Find resultant member forces (Pythagorean Formula)

Example:

Solve the following truss using the method of joints.

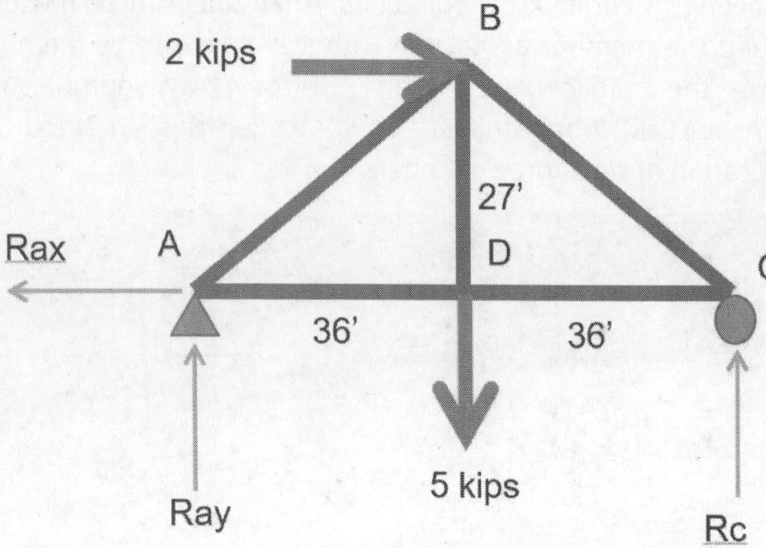

Step 1: Check for horizontal, vertical and rotational equilibrium

$\Sigma F_x = 0$ $\quad$ $2 - R_{ax} = 0$ $\quad$ $R_{ax} = 2$ kips

$\Sigma F_y = 0$ $\quad$ $R_{ay} + R_c - 5 = 0$ $\quad$ $R_{ay} + R_c = 5$

$\Sigma M = 0$

ΣM @A:

$-R_c \times (36+36) + 5 \times 36 + 2 \times 27 = 0$ $\quad$ $R_c = 3.25$ kips

$R_{ay} = 5 - 3.25 = 1.75$ kips

Step 2: Any zero-force member? No

Step 3: Solving joints by decomposing force vectors to their horizontal and vertical components

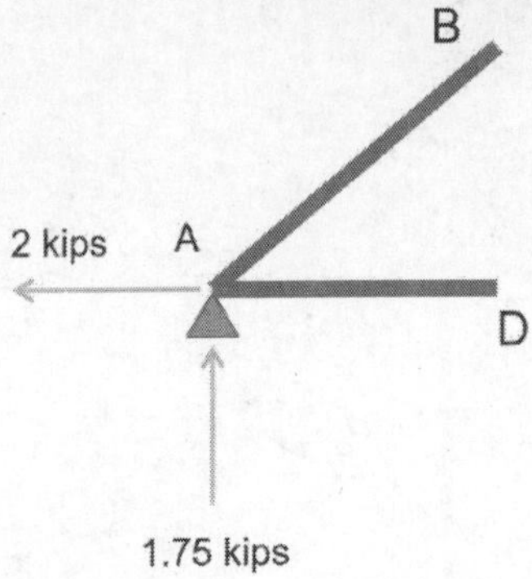

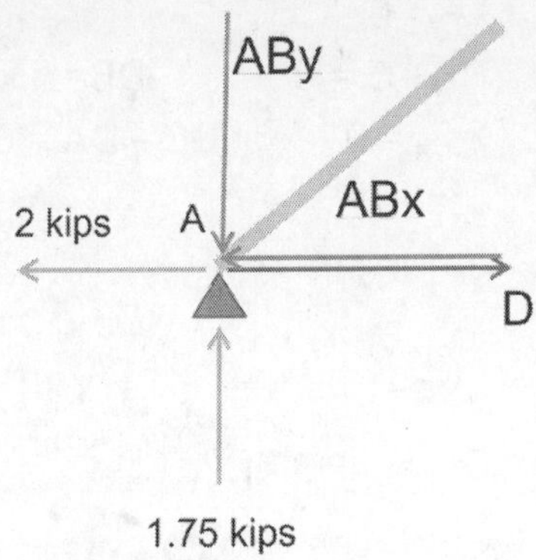

$\Sigma F_y = 0$

$-AB_y + 1.75 = 0$ $AB_y = 1.75$ downward

$\Sigma F_x = 0$ $Ab_x + AD - 2 = 0$

$AB_x / AB_y = 36/27$ $\Rightarrow Ab_x = Ab_y \times 36/27 = (-1.75) \times 36/27 = -2.33$ kips

$-2.33 + AD - 2 = 0$

AD= 4.33 kips it is toward the positive direction of x axis

$AB = \sqrt{(AB_x^2 + AB_y^2)} = \sqrt{(2.33)^2 + (1.75)^2} = 2.91$ kips

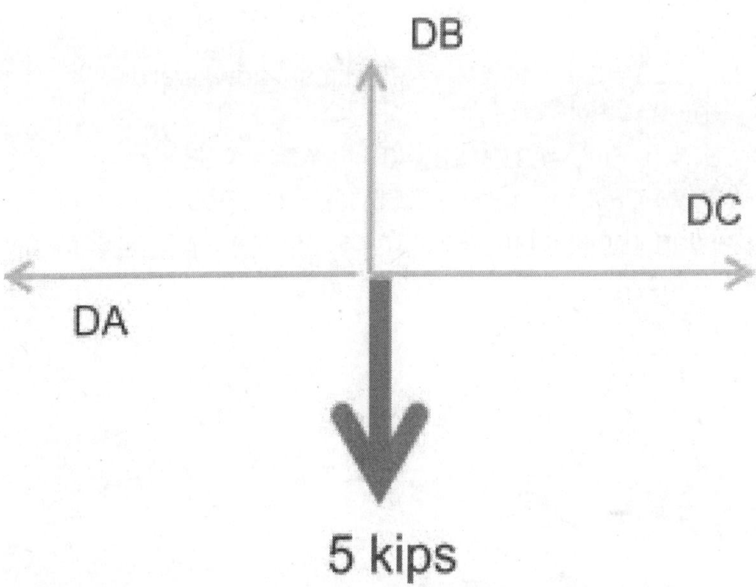

$\Sigma F_y = 0$

$DB - 5 = 0$ DB= 5 kips

$\Sigma F_x = 0$ $DC - DA = 0$ DC = 4.33 kips

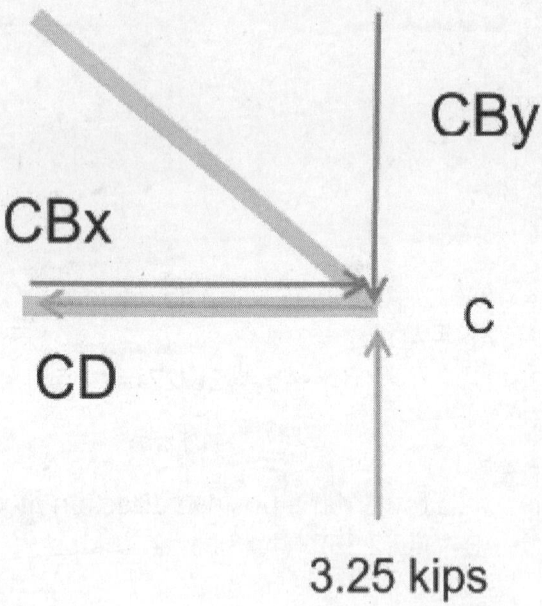

$\Sigma F_y = 0$

$- CB_y + 3.25 = 0$ $CB_y = 3.25$ kips downward

$\Sigma F_x = 0$

$-CD + CB_x = 0$ $-4.33 + CB_x = 0$ $CB_x = 4.33$ kips

$CB = \sqrt{(CB_x^2 + CB_y^2)} = \sqrt{(4.33)^2 + (3.25)^2} = 5.41$ kips

At the end, the axial forces in truss members are as shown in the diagram below:

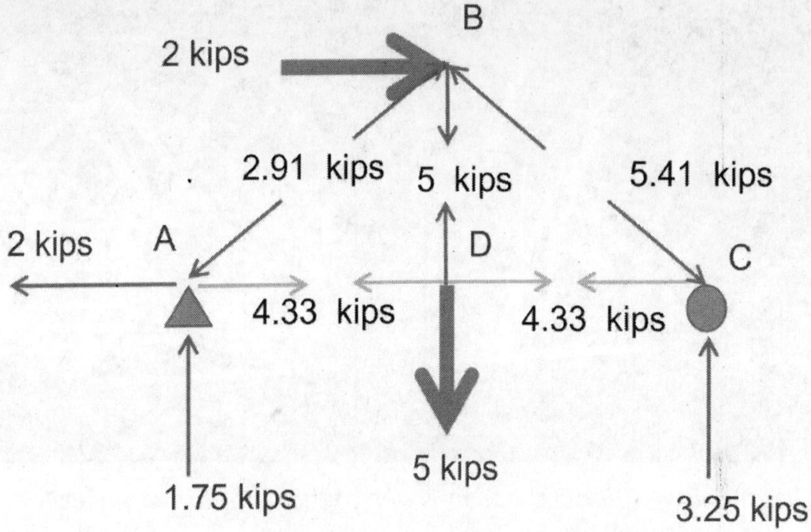

3D TRUSSES – SPACE FRAMES

Space frames are three-dimensional trusses where members are in tension and compression only. By definition, true trusses should have pinned connections while frames have rigid joints. Space frames may have both pinned and rigid connections. While the topology of space frames may be quite free of regular forms, the half-octahedron and tetrahedron are the common modules for creating a space frame structure.

Figure 5-7: A half-octahedron is one of the common modules for creating a space frame structure (Image source: space frame, 2002, wikipedia)

Figure 5-8: The free form of the space frame that covers the British museum courtyard ((Image source: British Museum, 2014, https://www.flickr.com/photos/22087304@N07/15629080951)

Figure 5-9: The pyramid of the Louvre entrance, designed by Chinese-American architect I. M. Pei, 1984 (Image source: Louvre Project (Pyramid), 1994, https://www.flickr.com/photos/69184488@N06/11876617365)

Space frames are relatively efficien and safe structural systems because even if a few members fail, the forces can reroute to remaining members.

RULE OF THUMB TO DETERMINE THE PROPORTIONS OF STRUCTURAL COMPONENTS

The economic depth-to-span ratio for steel trusses is 1:10 to 1:20, and for timber trusses is 1:6 to 1:10. The spacing of trusses in roof structures should be 20 to 30 ft for steel structures and 12 to 27 ft for timber trusses. The economic spans of different trusses are shown in the following table.

The suitable depth to span ration of space frames is usually between 1/10 to 1.20 of the span.

Table 5-1: The economic span ratio of trusses [3]

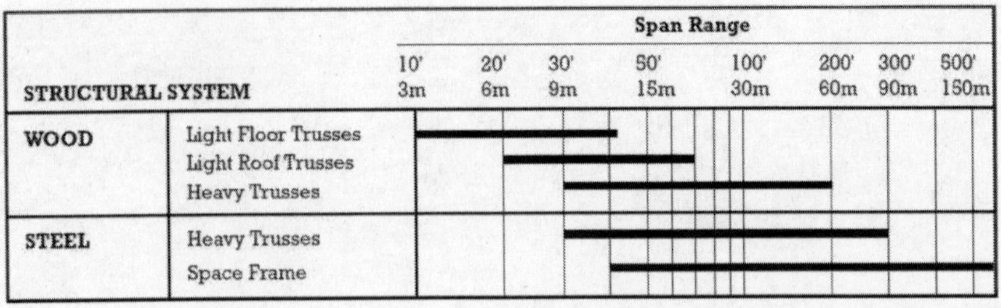

TENSEGRITIES

One or more interactive elements has been excluded from this version of the text. You can view them online here: https://pdx.pressbooks.pub/archistructures/?p=28#oembed-4

Video 5-3: What is a tensegrity structure (https://www.youtube.com/watch?v=BzgxYpDyO0M&t=1s)

A tensegrity structure is a stable three-dimensional space frame assembly of continuous cables and discontinuous struts where the struts do not touch one another. In tensegrity structures, tension elements stabilize the compression elements. One of the natural examples of tensegrity structures is the human body, where muscles (the tension elements) stabilize the compression elements (the bones).

Tensegrities were invented by sculptor Kenneth Snelson in 1948 and developed and patented by Buckminister Fuller. Snelson and Fuller's tensegrity theory translated into practice when David Geiger reduced the degree of indeterminacy.

TOPICS FOR CRITICAL THINKING

- The best way to understand how tensegrity structures work is to make a physical model. Let's get started by watching the following instructional video (https://www.youtube.com/watch?v=DQxNPhR20r0) on making a tensegrity model!

One or more interactive elements has been excluded from this version of the text. You can view them online here: https://pdx.pressbooks.pub/archistructures/?p=28#oembed-2

Video 5-4: How to make a tensegrity structure?

- Do you know any implications of tensegrity structures in the building industry?
- Watch the following video (https://www.youtube.com/watch?v=Rxtf5wHXkTA&list=PLKPD9oscWijEdwZFVuGbERolnw2auR1E6&t=778s) from its 11th minute to learn about how pioneers of designing tensegrity structures used symmetry in the design and analysis of spatial structures.

One or more interactive elements has been excluded from this version of the text. You can view them online here: https://pdx.pressbooks.pub/archistructures/?p=28#oembed-5

Video 5-5: Using symmetry in the design and analysis of spatial structures

- In the following video (https://www.youtube.com/watch?v=U2lKx_FYXWo&list=PLKPD9oscWijEdwZFVuGbERolnw2auR1E6), Professor Ken'ichi Kawaguchi, Architecture Professor at the University of Tokyo discusses the challenges of tensegrity construction. Move to the 11th minute to learn more about his findings

One or more interactive elements has been excluded from this version of the text. You can view them online here: https://pdx.pressbooks.pub/archistructures/?p=28#oembed-6

Video 5-6: Challenges of tensegrity construction

CHAPTER 6

Lattice Domes

Have you heard about lattice domes? Even if not, you may have heard of some outstanding projects, such as Eden Project, where a lightweight dome structure is designed to create a translucent space. Watch video 6-1 (https://www.youtube.com/watch?v=MWr67v620kY&feature=emb_logo) to learn about the design of compounded geodesic domes in the Eden Project.

One or more interactive elements has been excluded from this version of the text. You can view them online here: https://pdx.pressbooks.pub/archistructures/?p=30#oembed-1

Video 6-1: The geodesic dome of the Eden Project

A lattice dome consists of discrete and normally elongated elements shaping a grid pattern. Lattice domes may include a single layer or multiple layers, may have pinned or rigid joints, and may work only in tension and compression or in tension, compression, and bending. Lattice domes are named according to their grid patterns and the geometry of their curvature. Some examples of lattice domes are illustrated below:

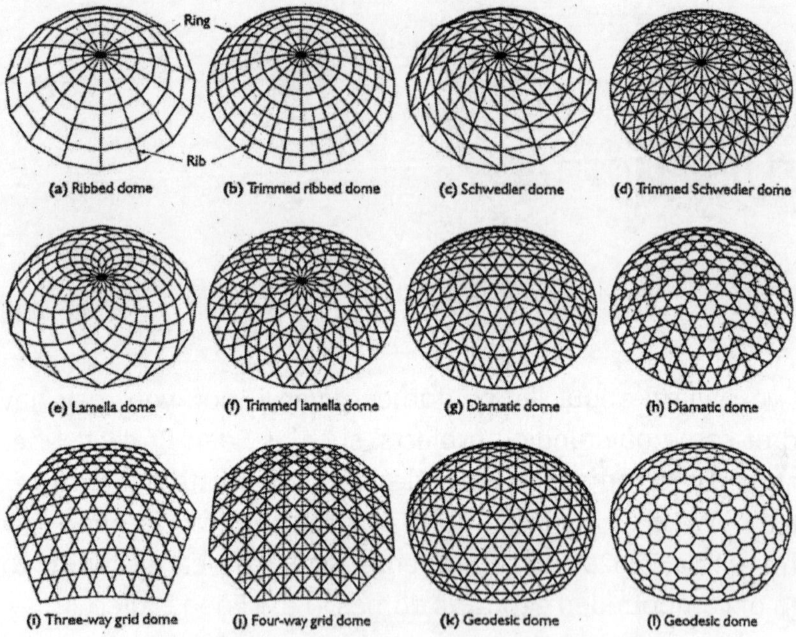

Figure 6-1: Different types of lattice domes (Image resource: [4])

Configuration processing of this type of spatial structure using the Euclidian geometric system is usually complicated and requires a great amount of time and effort. In the late 90s, Professor Hoshyar Nooshin, University of Surrey, Guilford, UK, developed a new geometric-algebraic system called Formex algebra. Formex algebra allows a designer to define the geometrical formulation of forms through concepts that affect movement, propagation, deformation, and curtailment. The creation of any type of spatial structure, such as space trusses, domes, vaults, hypar shells, polyhedric and free forms, can be carried out by using this mathematical system and its associated programming language, Formian. This new algebraic system, along with its designated software called Formian, allows convenient and accurate configuration processing of spatial structures. In 2018, he won the Torroja Medal – one of the highest honors in structural engineering – in recognition of his outstanding contribution to the field of spatial structures for over half a century.

Different types of lattice domes

Ribbed dome

A ribbed dome consists of several ribs and rings that intersect each other at rigid joints. A rib is entitled to a group of elements that lie along a meridional line, and a ring is entitled to a group of elements that constitute a horizontal polygon.

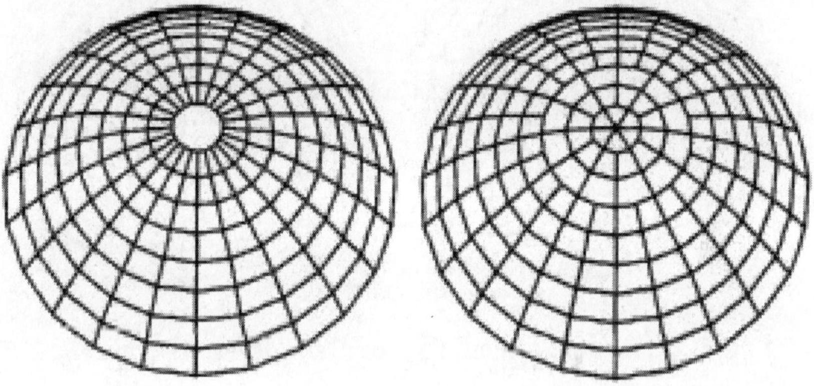

Figure 6-2: Examples of ribbed domes (Image source: [5])

Schwedler dome

A Schwedler dome includes intersecting ribs, rings, and diagonal elements. The name of this category of lattice domes is the namesake of the German Engineer J W Schwedler, who built several of this type in the 19th century. Sometimes due to the overcrowding of the elements near the crown of a ribbed or Schwedler dome, some elements are trimmed. In these cases, the dome will be called a trimmed ribbed dome or a trimmed Schwedler.

Figure 6-3: Some examples of Schwedler domes (Image source: [5])

Lamella dome

A lamella dome consists of intersecting diagonal elements and may or may not include one or more rings.

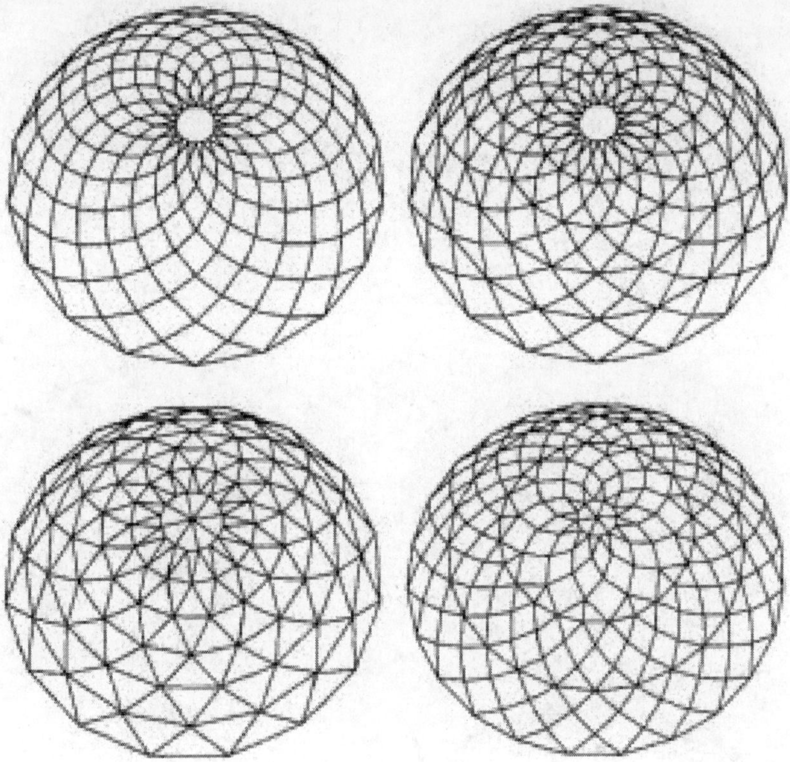

Figure 6-4: Examples of lamella domes (Image source: [5])

Diamatic dome

Diamatic domes consist of several sectors whose side boundaries are along two meridians of the circumsphere of the dome, and the bottom boundary is along a parallel of the circumsphere. An important characteristic of diamatic domes is that they do not present any problem regarding 'element cluttering' near the crown.

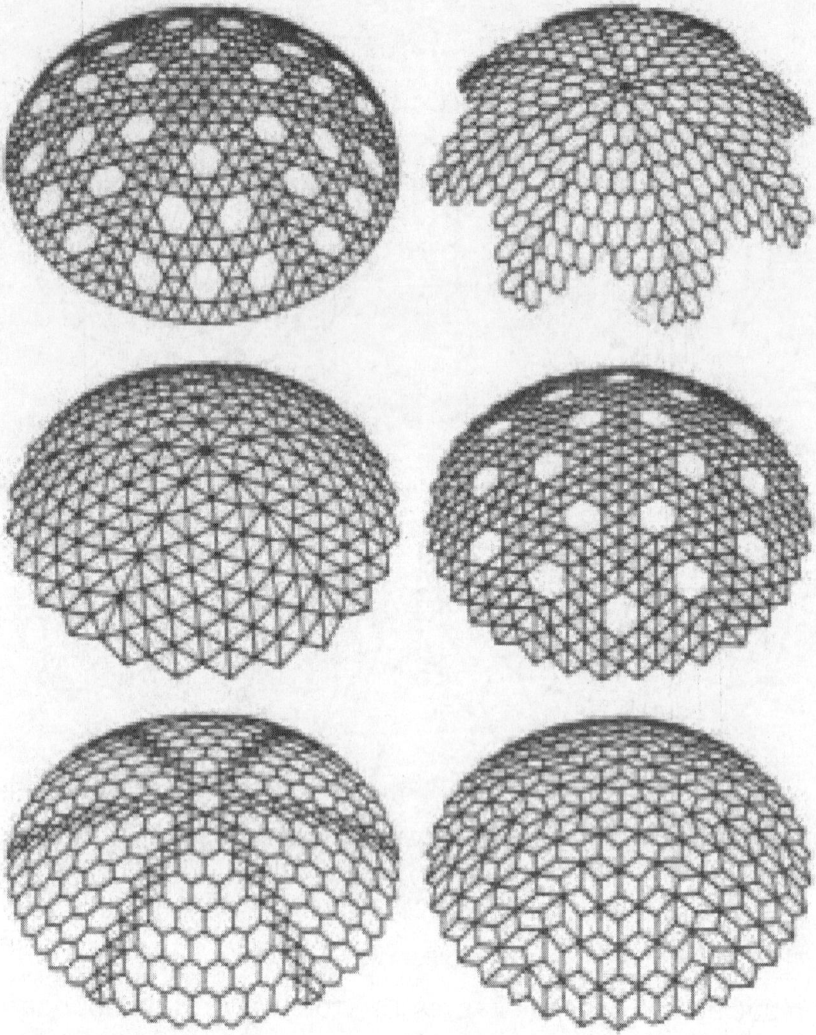

Figure 6-5: Some examples of diamatic domes (Image source: [6])

One of the well-known examples of a diamatic dome is the Tacoma dome. Tacoma wooden dome is a multi-purpose arena in Tacoma, Washington (1983). The arena has a 530-ft diameter and 152-ft height.

Figure 6-6: Tacoma dome, Washington, US (Image source: The Tacoma Dome, seen from the Bridge of Glass, 2006, wikipedia)

Scallop domes

A scallop dome, similar to the marine creature scallop whose shell has arched ribs, consists of arched sectors that are separately arched. The dome itself had a general curvature, and the operation of scalloping further improves the structural behavior of the dome.

Figure 6-7: Examples of sea shells that inspire the configuration of scallop domes (Image source: Small sea shell, 2018, https://commons.wikimedia.org/wiki/File:Small_sea_shell.jpg)

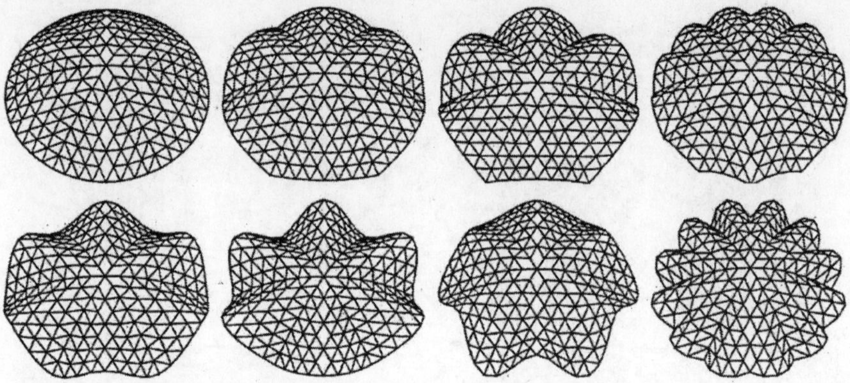

Figure 6-8: Examples of scallop domes (Image source: [8])

Conical domes

The geometry of a conical dome simply relies on the geometry of a cone and may be created by mapping different patterns onto its surface.

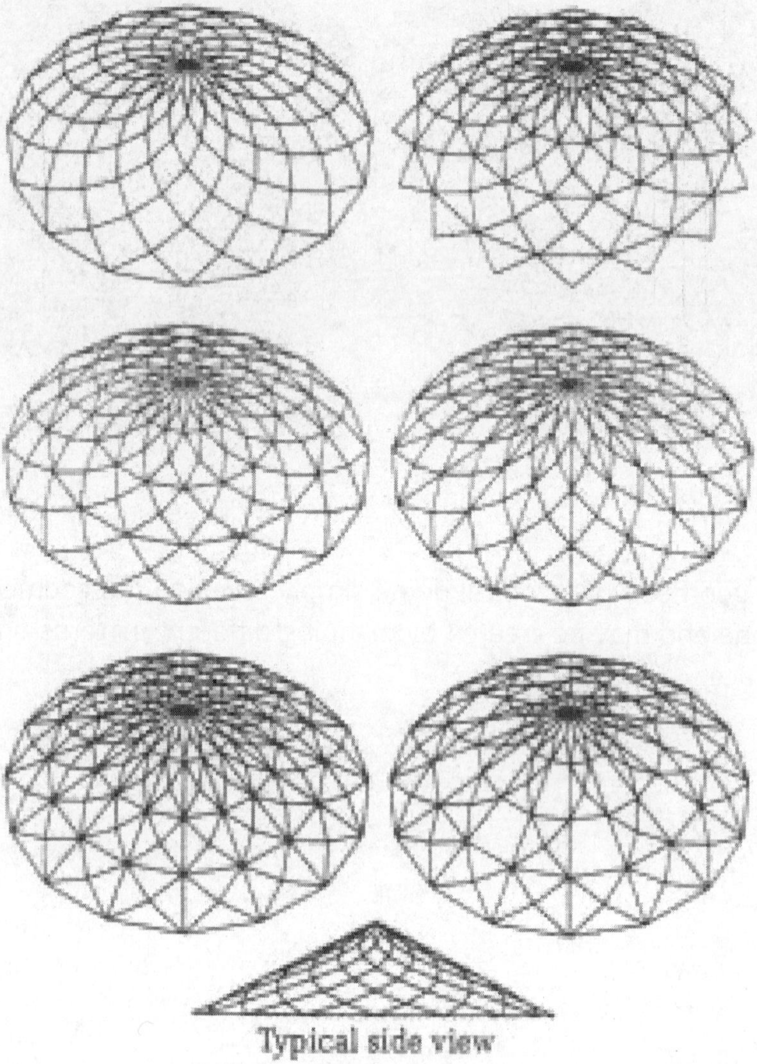

Figure 6-9: Examples of conical domes (Image source: [7])

Mallow domes

Mallow domes, so-called because they resemble the mallow flower, consist of a number of sectors with a hyperbolic paraboloidal grid.

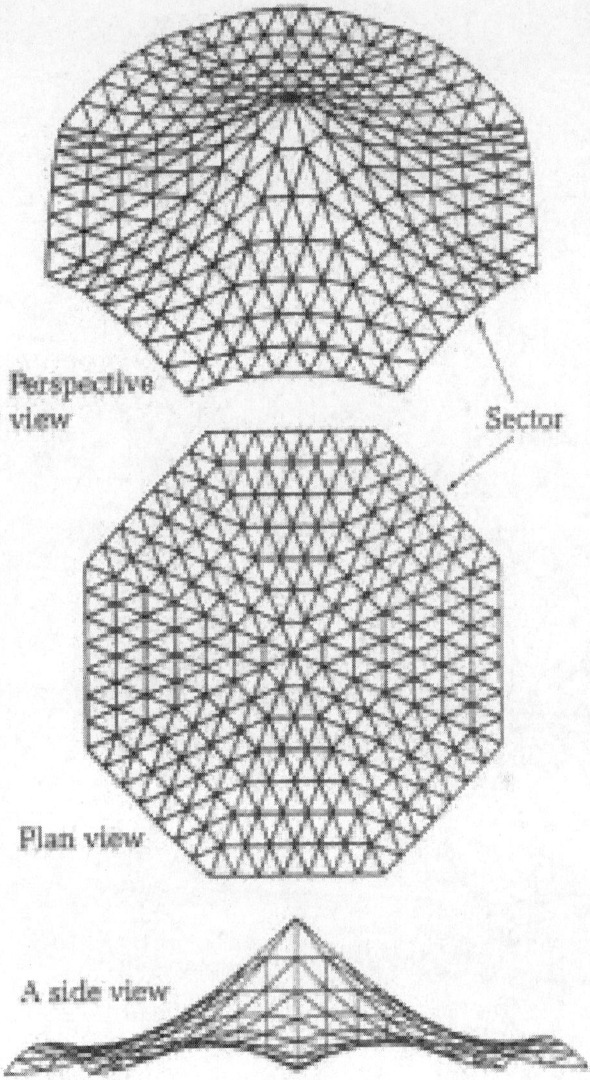

Figure 6-10: An example of a mallow dome (Image source: [7])

Onion domes

Onion domes have a bulb-shaped body with a pointed crown and may be created with different grid patterns.

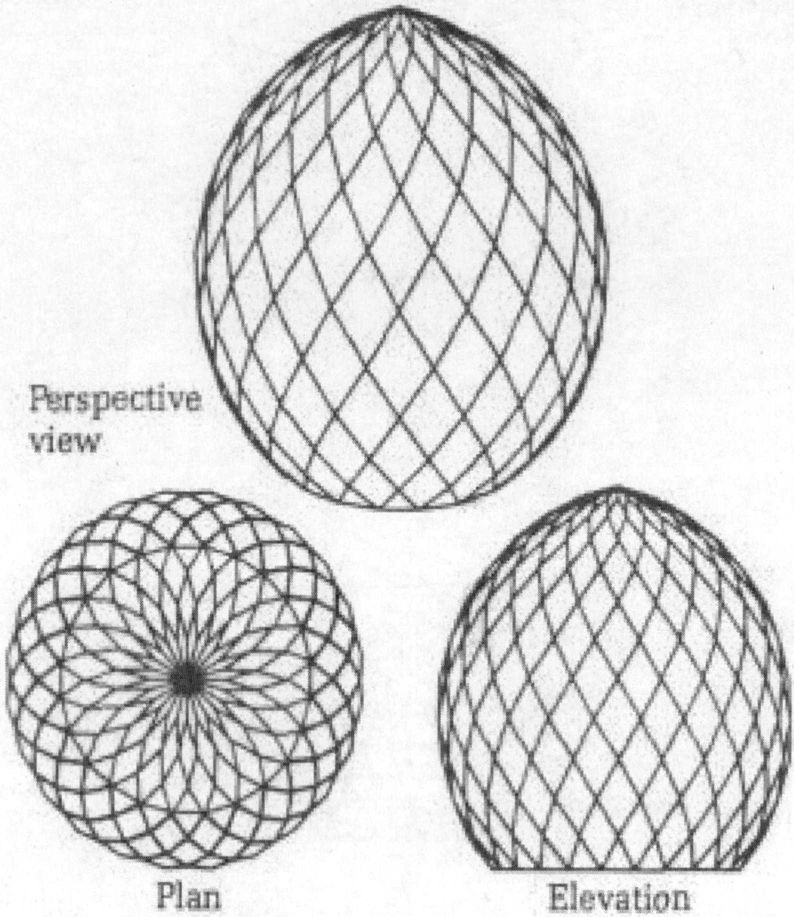

Figure 6-11: An example of an onion dome (Image source: [7])

Geodesic domes

A geodesic dome is created by mapping a triangulated pattern onto a number of faces of a polyhedron and projecting the results onto a sphere that is concentric with the initial polyhedron. An icosahedron (20-faced regular polyhedron) is one of the common volumes used in creating geodesic domes.

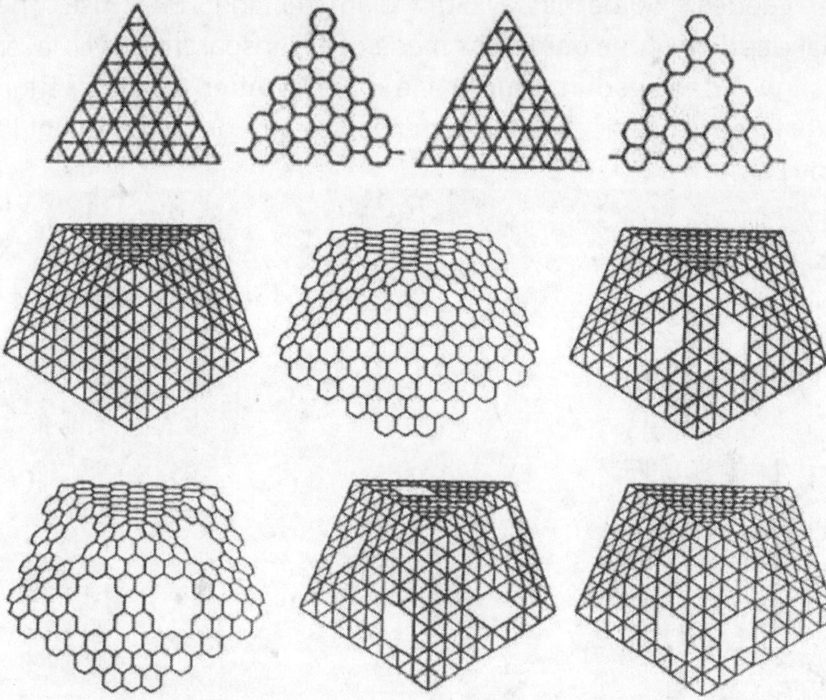

Figure 6-12: Mapping a pattern on the faces of a polyhedron (Image source: [9])

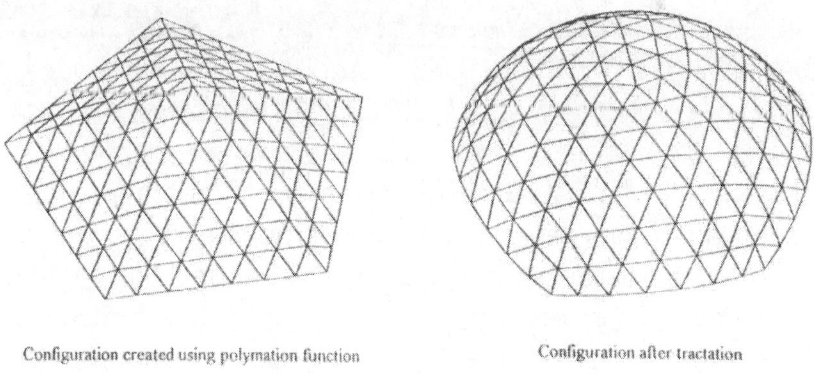

Figure 6-13: Mapping an icosahedron onto a hemisphere to gain a geodesic dome (Image source: [9])

One famous geodesic dome is the US Pavilion in Expo 67, designed by Buckminster Fuller and built in Montreal. The double-layer geodesic dome has a 250-ft diameter and 200-ft rise. The initial cladding of the dome was made of a transparent acrylic layer which was damaged in a major file in 1977. After the renovation in 1994, the geodesic skeleton remained as an iconic monument in Montreal.

Figure 6-14: Bio dome, by Buckminster Fuller, Montreal (Image source: Biosphere Montreal, 2007, Wikimedia, https://commons.wikimedia.org/wiki/File:Biosphere_montreal.JPG)

Missouri Botanical Gardens Climaton is another example of geodesic domes designed and built in double layers. The geodesic dome is a quarter sphere spanning 175 ft and housing the plant collection of the Missouri botanical garden. The dome is made of aluminum tubes and steel cables. The rise of the dome is 70 ft, and current cladding of the dome consists of freestanding glass panels.

Figure 6-15: Missouri Botanical Gardens Climaton (Image source: The Climatron greenhouse at the Missouri Botanical Garden, 2004, https://en.wikipedia.org/wiki/Climatron#/media/ File:Climatron,_Missouri_Botanical_Gardens.jpg)

TOPICS FOR CRITICAL THINKING

- Watch the following video (https://www.youtube.com/watch?v=6a9ffR_tfd4&list=PLKPD9oscWijEdwZFVuGbERolw2auR1E6&t=2s) from timestamp 8:12 to learn about the design of an innovative lattice dome in the Dutch Marine Museum, Amsterdam, the Netherlands.

One or more interactive elements has been excluded from this version of the text. You can view them online here: https://pdx.pressbooks.pub/archistructures/?p=30#oembed-2

Video 6-2: The design of an innovative lattice dome in the Dutch Marine Museum

- Are you interested in building a geodesic dome of your own? Watch this video (https://www.youtube.com/watch?time_continue=1&v=ryAGFPobYUY&feature=emb_logo) and find a full list of videos on designing and assembling a geodesic dome.

One or more interactive elements has been excluded from this version of the text. You can view them online here: https://pdx.pressbooks.pub/archistructures/?p=30#oembed-3

Video 6-3: Design and construction of a geodesic dome

CHAPTER 7

Material Properties

In the previous chapters, we have discussed the determination of internal forces produced in structural members by externally applied loads. In chapters 7 and 8, we will discuss the relationship between loads applied to a non-rigid body and the resulting internal forces and deformations induced in the body. These analyses will be used to determine the required size of a structural element to safely resist external loads.

STRESS AND STRAIN

Stress is a term that is used to describe the intensity of a force. By definition, stress is force per unit area of a material and may be described in psi (pounds per square inch), psf (pounds per square foot), MN/m^2 (Meganewton per square meter), kgf/cm^2 (kilogram-force per square centimeter), MPa (megapascal = N/mm^2). Therefore, the axial stress can be represented mathematically via the following equation:

$$\sigma = \frac{P}{A}$$

Where:

Stress = σ = load/area = P/A

P = tensile/compression load or force
A = the area

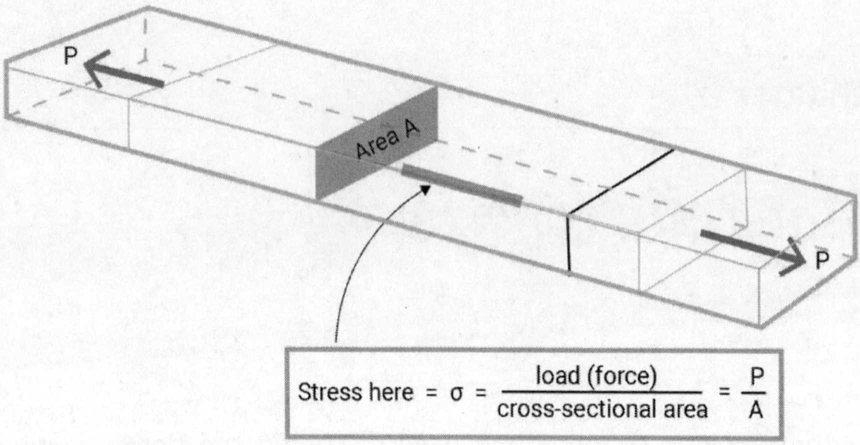

Figure 7-1: Stress is force per unit area of a material

Any structural member under load has some amount of deformation. The member may be compressed or stretched due to a compression or tension force. Some materials such as steel do deform as well, even though their deformation is not significant. We measure this deformation per unit length and call it strain.

$$\varepsilon = \frac{D}{L}$$

Where:

Strain = ε= deformation/original length = D/L
D = deformation
L = original length

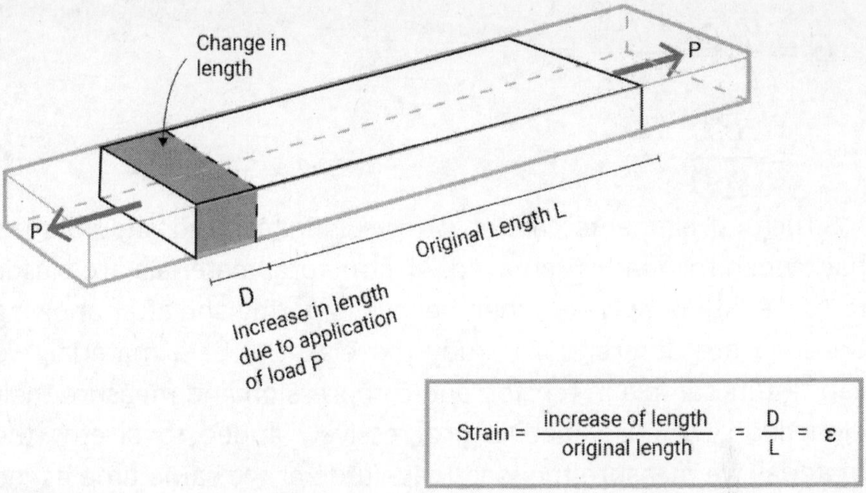

Figure 7-2: Any structural member under load has some amount of deformation or change of shape. Strain is the deformation per unit length

MODULUS OF ELASTICITY

When a structural element is stretched on a molecular scale, the molecules are pulled apart. The electromagnetic bond among the molecules resists against the externally applied force and creates the elastic behavior of the element. In the late 17th century, Robert Hooke was the first fellow who discovered that there is a relation between the stress in an element and the developed strain. However, this relationship was formulated by Thomas Young in the 19th century. The relationship between stress and strain is a measure of the elasticity of the material and is called Young's Modulus or the Modulus of Elasticity (E).

$$E = \frac{\sigma}{\varepsilon}$$

Where:

E= Modulus of Elasticity

ε= Strain= deformation/original length = D/L

σ = Stress = load/area = P/A

$$E = \frac{\sigma}{\varepsilon} = \frac{P/A}{D/L} =$$
$$E = \frac{PL}{AD} =$$

Structural elements cannot deform infinitely and should snap back when the load is removed. All structural materials are elastic to some extent; however, their behaviors during and after applying the load are different. To study the elasticity of a material, we can test its sample in tension and compression and measure their elongation. When a load is progressively added to stretch the material, we measure the load magnitude at the same time as the amount the sample stretches.

One or more interactive elements has been excluded from this version of the text. You can view them online here: https://pdx.pressbooks.pub/archistructures/?p=32#oembed-1

Video 7-1: Material properties and fundamental terms used to define them (https://www.youtube.com/watch?v=BHZALtqAjeM)

During this test, you can see that there is a stress limit before which the material returns to its original shape if the load is removed. But beyond that stress limit, the material does not snap back, and the deformation will remain permanent. This stress limit is called the yield stress, and the corresponding elongation is called inelastic or plastic deformation. Then, there is a stress limit, called ultimate stress, where the maximum load can be placed before the breaking of the element. Finally, beyond the ultimate stress, the element deforms rapidly and breaks at the point of fracture.

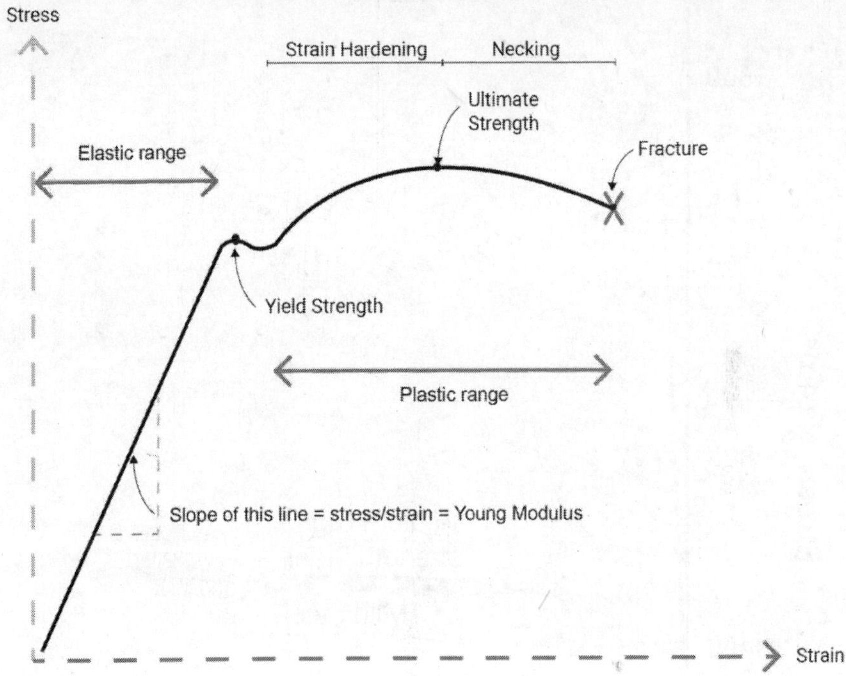

Figure 7-3: A stress-strain plot of a material

The stress-strain diagram is unique for any given material. Young's modulus represents a measure of the stiffness of the material. Thus, the slope of the straight line varies greatly for different materials. When a material is referred to as "flexible," this just means a material has a low Young's modulus. Moreover, when some materials exceed their yield point, they may crack or suddenly break into pieces. They may not have any plastic deformation, and or their plastic range is remarkably limited. These materials are brittle. In contrast, some materials such as low-carbon steel, aluminum, or copper have a significant plastic range that can warn of impending failure. This type of material behavior is termed ductile.

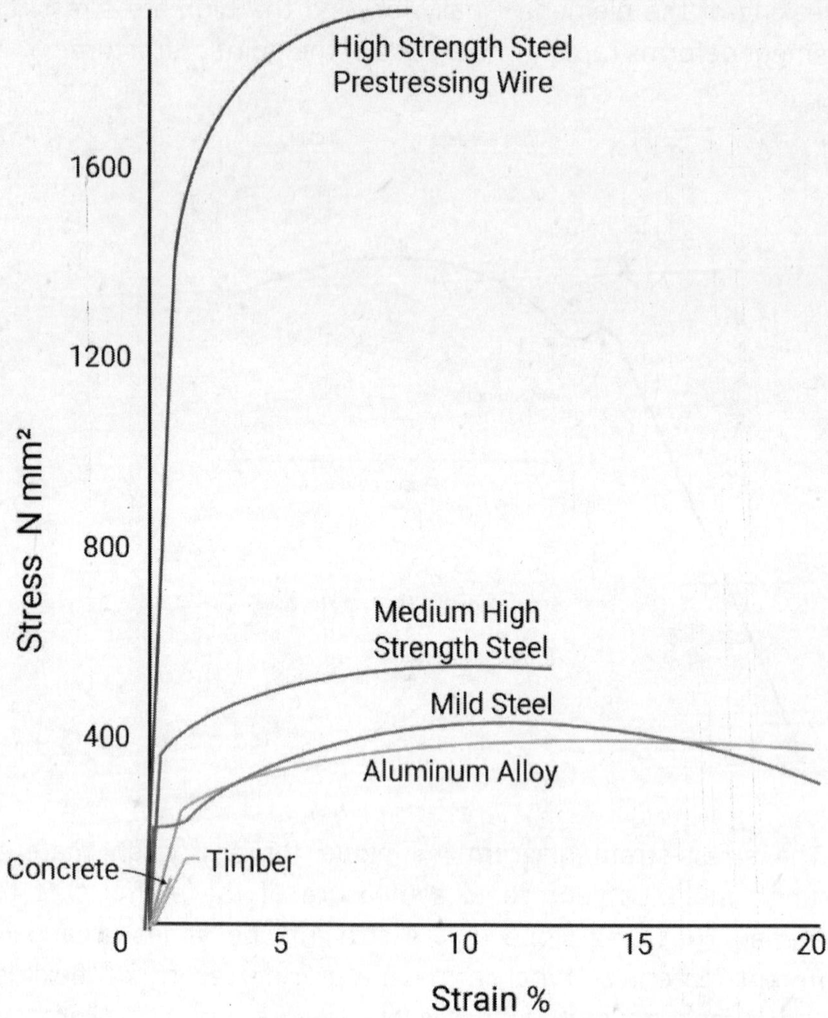

Figure 7-4: Using stress-strain graph to compare properties of different materials

One or more interactive elements has been excluded from this version of the text. You can view them online here: https://pdx.pressbooks.pub/archistructures/?p=32#oembed-2

Video 7-2: Understanding Young's modulus (https://www.youtube.com/watch?v=DLE-ieOVFjI&feature=emb_logo)

Example

What is the minimum required cross-sectional area for the two columns supporting the beam shown below?

Allowable stress:
Wood = 800 psi
Steel = 21 ksi
Concrete = 3000 psi
Clay Soil = 2000 psf

* Allowable stress is the maximum stress (tensile, compressive, or bending) that is allowed to be applied on a structural material. Building codes generally define the allowable stresses, which for steel and aluminum are fractions of their yield stress.

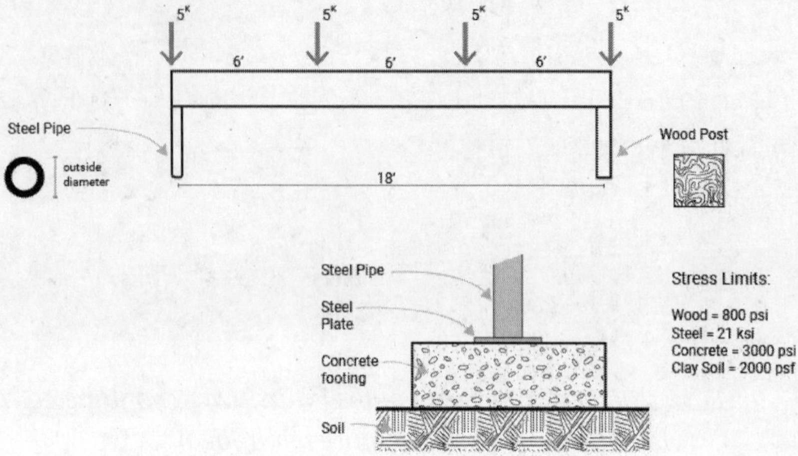

$\Sigma F_x = 0$

$\Sigma F_y = 0$ All vertical forces must balance out to zero.

$R_1+R_2-5-5-5-5=0$

$R_1+R_2=20$

We have a symmetric loading condition. So $R_1=R_2$

$R_1=R_2 = 20/2 = 10$ k

Allowable stress: $\sigma = \frac{P}{A}$

For wood column :

$800 = 10,000 / A \Rightarrow A = 10,000 / 800 = 12.5 \text{ in}^2$ of wood

Each side should be greater than $\sqrt{12.5}=3.54$

For steel column

$21,000 = 10,000 / A \Rightarrow A = 10,000 / 21,000 = 0.48 \text{ in}^2$ of steel

Concrete Allowable stress $\sigma = P / A$

$3000 = 10,000 / A \Rightarrow A = 10,000 / 3000 = 3.33 \text{ in}^2$ steel base plate

Soil Allowable stress $\sigma = P / A$

$2000 = 10,000 / A \Rightarrow A = 10,000 / 2,000 = 5 \text{ ft}^2$ of footing surface

TOPICS FOR CRITICAL THINKING

- How can we use Allowable stress: σ =\frac{P}{A} for calculating the shear stress in structural elements?
- Test a rubber band and find its Young's Modulus. You may watch this video (https://www.youtube.com/watch?v=aGS_tYML3HQ) to learn about similar experiments.

One or more interactive elements has been excluded from this version of the text. You can view them online here: https://pdx.pressbooks.pub/archistructures/?p=32#oembed-3

Video 7-3: Determining Young's modulus of a material

CHAPTER 8

Cross-sectional Properties

Designing Structural elements requires the knowledge of applied external loads and internal reactions, material strengths, and cross-sectional properties. The geometrical properties of a structural element are critical in keeping axial, shear, and bending stresses within allowable limits and moderating the amount of deflection. The following demonstrations show how the shape of the cross-sections affects their stiffness

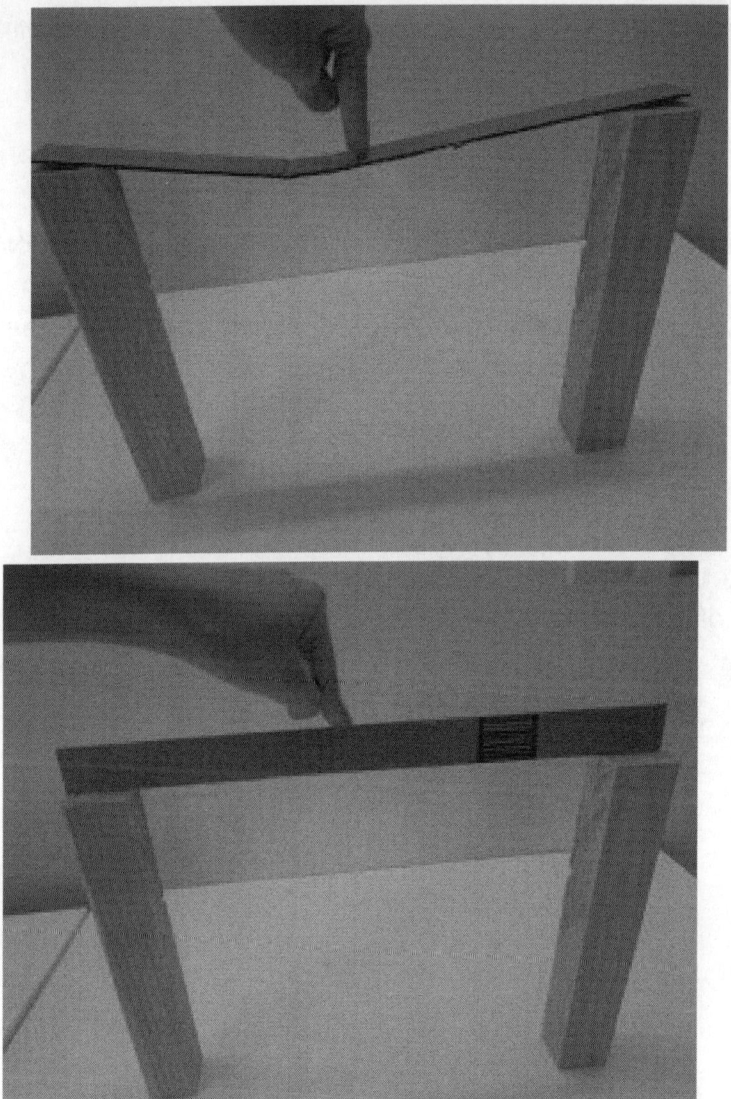

Figure 8-1: Relation between the cross-sectional depth (rise) of a beam and its stiffness

Chapter 7 discussed that "area" is one of the cross-sectional properties, which is important to reduce the amount of stress in beams and columns. This chapter focuses on the shape of the cross-sectional area and its distribution about the neutral axis of beams. In the following, some cross-sectional properties that will be necessary to calculate beams and columns are introduced.

CENTER OF GRAVITY

The center of gravity of a body is the point about which the mass of the body is balanced or the point through which the weight of the body acts. When the density of a body is uniform throughout, the center of gravity and the centroid (geometric center) of the body are at the same point.

- The centroid of a rectangle is defined as the center point where all the diagonals intersect each other.
- A centroid of a triangle is the point where the three medians of the triangle meet.

Shape	Drawing	$\overline{x}$	$\overline{y}$	Area
Rectangle	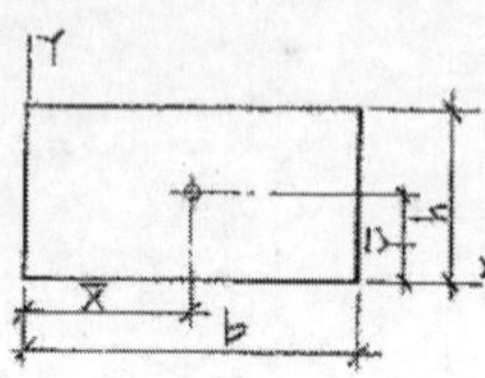	$b/2$	$h/2$	bh
Triangle	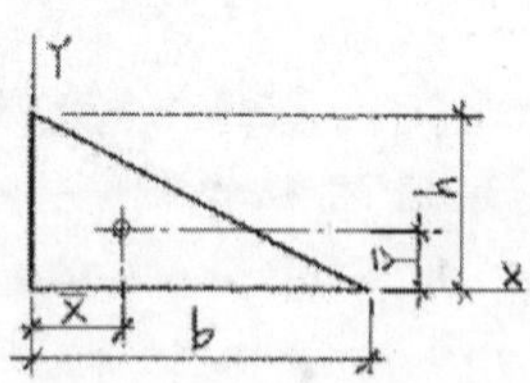	$b/3$	$h/3$	$bh/2$

Figure 8-2: The center of gravity of some basic geometrical shapes

Finding the center of gravity of the cross-section of a beam or column can define the location of neutral axis of that body. In a typical beam loaded by self-weight, every point above the neutral axis is in compression and every point below the neutral axis is in tension. At neutral axis the values of tension and compression is equal to zero.

One or more interactive elements has been excluded from this version of the text. You can view them online here: https://pdx.pressbooks.pub/archistructures/?p=34#oembed-1

Video 8-1: Neutral axis in a beam (https://www.youtube.com/watch?v=BthnS6LJt8s&t=1s)

You can use the following equation to find the center of Gravity of a compounded shape:

$$\bar{X} = \frac{\sum(Area \times d_x)}{\sum Area}$$

$$\bar{Y} = \frac{\sum(Area \times d_y)}{\sum Area}$$

Example

1ST MOMENT OF AREA

One or more interactive elements has been excluded from this version of the text. You can view them online here: https://pdx.pressbooks.pub/archistructures/?p=34#oembed-2

Video 8-2: A demonstration of the moment of areas of two rods (https://www.youtube.com/watch?v=m9weJfoW5J0)

By definition, the tendency of an area alone to rotate about an axis in the plane of that area.

$Q = A\bar{x}$

At the Neutral/Centroid axis:

$A_1 x_1 = A_2 x_2$

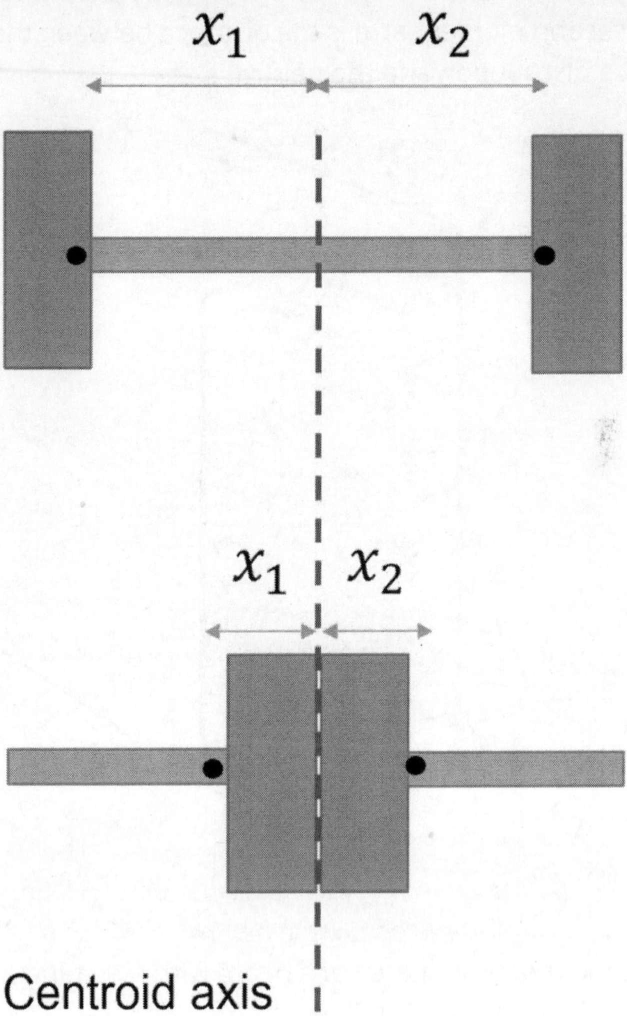

Figure 8-3: The difference of 1st moment of area in two beams with different cross sections

2ND MOMENT OF AREA/INERTIA

By definition, the 2nd moment of area is the distance of force distribution from the neutral axis. The 2nd moment of area involved the first moment of area multiply by a second moment arm. The second moment arm is the distance between the centroid of the force distribution and the neutral axis.

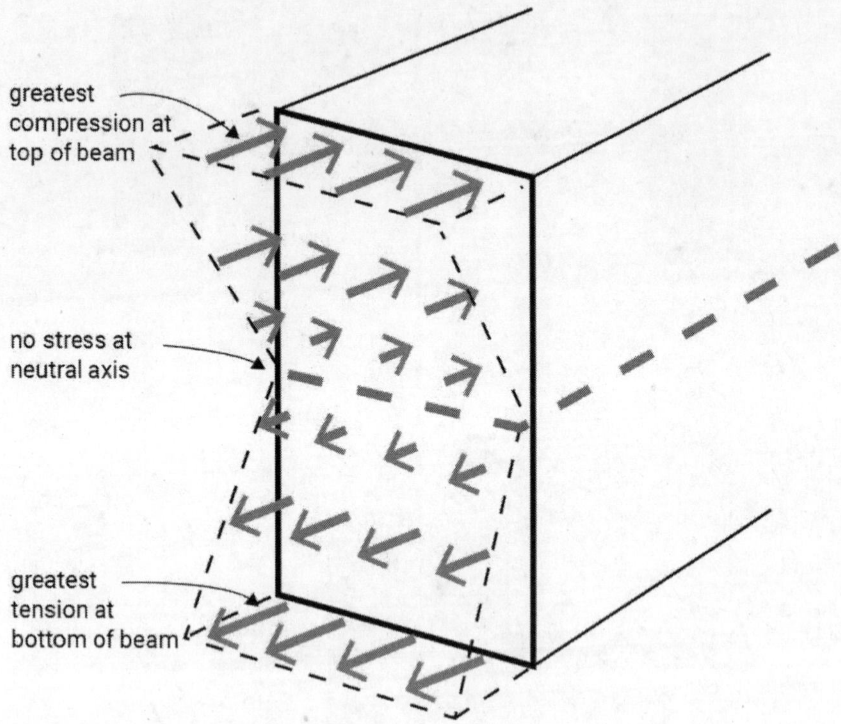

Figure 8-4: The pattern of force distribution in relation to the distance from the neutral axis

The second moment of area of a beam with a rectangular section can be calculated using the following equation:

$I_x = \dfrac{bh^3}{12}$

Where:

I_x = Second moment of area

h = depth of the beam

b = with of the beam

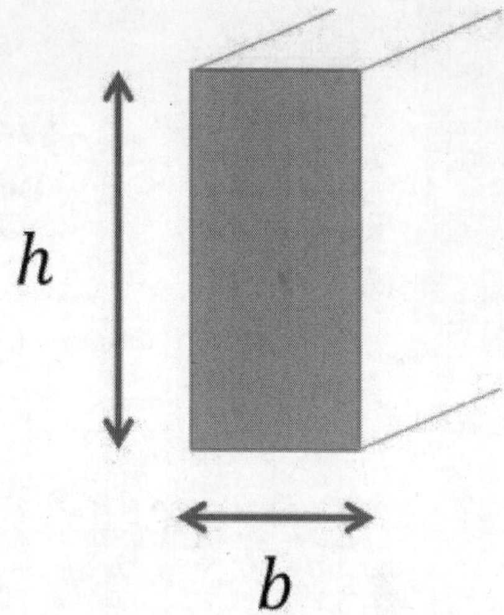

SECTION MODULUS

By definition, the section modulus (S_x) of a beam with a symmetric section equals its second moment of area divided by half its depth at the extreme fiber.

$S_x = \dfrac{I_x}{c}$

Where:

S_x = Section modulus

I_x = Second moment of area

c= h/2 at extreme fibers of a symmetric section

h = depth of the beam

The section modulus will help determine the cross-section shape of a beam as discussed in the Chapter 9.

GEOMETRICAL PROPERTIES OF STEEL BEAM CROSS-SECTIONS

Tables of design dimensions, detailing dimensions, axial flexure, strong-axis flexure, and weak-axis flexure of steel beams are provided in the Steel Construction Manual published by the American Institute of Steel Construction (AISC). You may find the cross-sectional area (A), depth of the beam (d), I_x, and S_x.

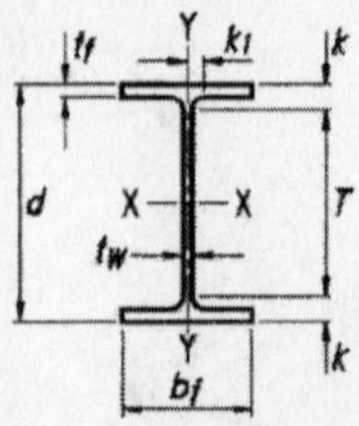

W-Shapes
Dimensions

Shape	Area, A	Depth, d		Web			Flange				Distance				
				Thickness, t_w		$\frac{t_w}{2}$	Width, b_f		Thickness, t_f		k		k_1	T	Workable Gage
											k_{des}	k_{det}			
	in.^{2}	in.		in.		in.	in.		in.		in.	in.	in.	in.	in.
W30×391[h]	115	33.2	$33\frac{1}{4}$	1.36	$1\frac{3}{8}$	$\frac{11}{16}$	15.6	$15\frac{5}{8}$	2.44	$2\frac{7}{16}$	3.23	$3\frac{3}{8}$	$1\frac{1}{2}$	$26\frac{1}{2}$	$5\frac{1}{2}$
×357[h]	105	32.8	$32\frac{3}{4}$	1.24	$1\frac{1}{4}$	$\frac{5}{8}$	15.5	$15\frac{1}{2}$	2.24	$2\frac{1}{4}$	3.03	$3\frac{1}{8}$	$1\frac{7}{16}$	↓	↓
×326[h]	95.9	32.4	$32\frac{3}{8}$	1.14	$1\frac{1}{8}$	$\frac{9}{16}$	15.4	$15\frac{3}{8}$	2.05	$2\frac{1}{16}$	2.84	$2\frac{15}{16}$	$1\frac{3}{8}$		
×292	86.0	32.0	32	1.02	1	$\frac{1}{2}$	15.3	$15\frac{1}{4}$	1.85	$1\frac{7}{8}$	2.64	$2\frac{3}{4}$	$1\frac{5}{16}$		
×261	77.0	31.6	$31\frac{5}{8}$	0.930	$\frac{15}{16}$	$\frac{1}{2}$	15.2	$15\frac{1}{8}$	1.65	$1\frac{5}{8}$	2.44	$2\frac{9}{16}$	$1\frac{5}{16}$		
×235	69.3	31.3	$31\frac{1}{4}$	0.830	$\frac{13}{16}$	$\frac{7}{16}$	15.1	15	1.50	$1\frac{1}{2}$	2.29	$2\frac{3}{8}$	$1\frac{1}{4}$		
×211	62.3	30.9	31	0.775	$\frac{3}{4}$	$\frac{3}{8}$	15.1	$15\frac{1}{8}$	1.32	$1\frac{5}{16}$	2.10	$2\frac{1}{4}$	$1\frac{3}{16}$		
×191[c]	56.1	30.7	$30\frac{5}{8}$	0.710	$\frac{11}{16}$	$\frac{3}{8}$	15.0	15	1.19	$1\frac{3}{16}$	1.97	$2\frac{1}{16}$	$1\frac{3}{16}$		
×173[c]	50.9	30.4	$30\frac{1}{2}$	0.655	$\frac{5}{8}$	$\frac{5}{16}$	15.0	15	1.07	$1\frac{1}{16}$	1.85	2	$1\frac{1}{8}$	↓	↓

W-Shapes
Properties

W30-W27

Nominal Wt.	Compact Section Criteria		Axis X-X				Axis Y-Y				r_{ts}	h_o	Torsional Properties	
	$\frac{b_f}{2t_f}$	$\frac{h}{t_w}$	I	S	r	Z	I	S	r	Z			J	C_w
lb/ft			in.^{4}	in.^{3}	in.	in.^{3}	in.^{4}	in.^{3}	in.	in.^{3}	in.	in.	in.^{4}	in.^{6}
391	3.19	19.7	20700	1250	13.4	1450	1550	198	3.67	310	4.37	30.8	173	366000
357	3.45	21.6	18700	1140	13.3	1320	1390	179	3.64	279	4.31	30.6	134	324000
326	3.75	23.4	16800	1040	13.2	1190	1240	162	3.60	252	4.26	30.4	103	287000
292	4.12	26.2	14900	930	13.2	1060	1100	144	3.58	223	4.22	30.2	75.2	250000
261	4.59	28.7	13100	829	13.1	943	959	127	3.53	196	4.16	30.0	54.1	215000
235	5.02	32.2	11700	748	13.0	847	855	114	3.51	175	4.13	29.8	40.3	190000
211	5.74	34.5	10300	665	12.9	751	757	100	3.49	155	4.11	29.6	28.4	166000
191	6.35	37.7	9200	600	12.8	675	673	89.5	3.46	138	4.06	29.5	21.0	146000
173	7.04	40.8	8230	541	12.7	607	598	79.8	3.42	123	4.03	29.3	15.6	129000

Figure 8-5: Example of steel profiles listed in Steel Construction Manual [15]

CHAPTER 9

Shear and Bending Stress in Simple Beams

Before discussing shear and bending stress in simple beams, let's watch video 9-1 (https://www.youtube.com/watch?v=SZM0kGBote4&t=1s) and review what the role of beams and columns are in a structural system and how they generally behave under dead and live loads. At the end of this video an online beam simulator is introduced that you can explore it further here.

One or more interactive elements has been excluded from this version of the text. You can view them online here: https://pdx.pressbooks.pub/archistructures/?p=36#oembed-1

Video 9-1: General behavior of beams and columns in a structural system

SHEAR AND BENDING FORCES IN SIMPLE BEAMS

As we discussed in the previous chapters, the level of stress in a structural element depends on the applied external loads as well as its surface area or cross-sectional properties. Before addressing the shear and bending stress in beams, let's overview the maximum shear and bending loads in a simple beam. The two diagrams below show the end reactions, maximum values of the shear load, and the bending moment in a simple beam supported by a pinned joint and a roller. The first diagram shows the maximum values for a beam under a point load, and the second shows the respective amounts for a beam under a uniformly distributed load.

Maximums in a simple beam under a point load:

End reaction = P/2

Shear load = P/2

Bending moment = PL/4

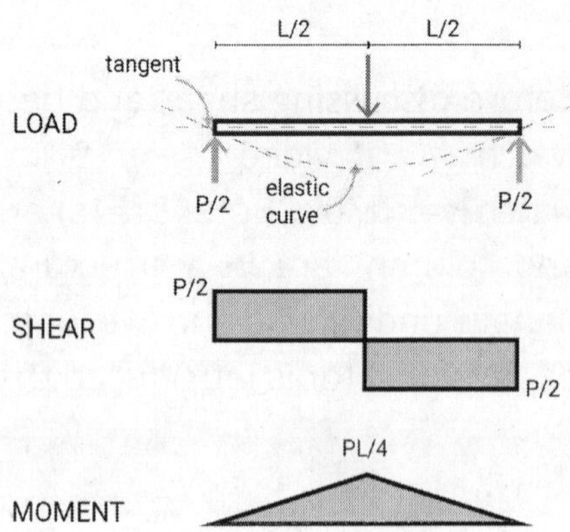

Maximums in a simple beam under a uniformly distributed load:

Equivalent point-Load = wL

End reaction $R_1=R_2=wL/2$

Shear load $V_{max}=wL/2$

Bending moment $M_{max}=\dfrac{(wL)^2}{8}$

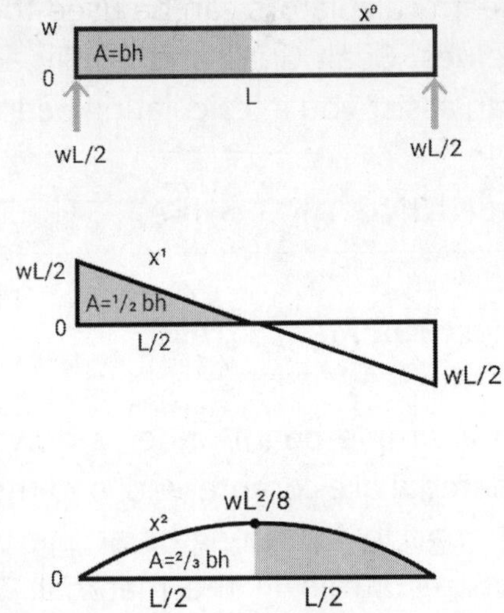

Figure 9-1: The end reactions, maximum values of the shear load, and the bending moment in a simple beam supported by a pinned joint and a roller

Sign conventions

As you can see, in the two diagrams, there are sign conventions for demonstrating the moment and shear in beams.

Sign Convention for Moment:

+ the top fibers in compression have positive curvature (holds water)

– the top fibers in tension have negative curvature (spills water)

Sign Convention for Shear:

+ the sum of the vertical forces to the left of the cut is upwards

– the sum of the vertical forces to the left of the cut is downwards

Maximum shear and Bending forces in different types of beams can be obtained by drawing a free-body diagram or referring to the AISC Steel Construction Manual and using the provided tables showing the shear and bending diagrams. Furthermore, online free beam calculators can be used to get maximum shear and bending values. BeamGuru and SkySiv are two free online platforms that can assist you in calculating beams.

BENDING AND SHEAR STRESS IN BEAMS

Elastic bending stress

In a simple beam under a downward load, the top fibers of the material are compressed, and the bottom fibers are stretched. The change in fiber lengths at the top and the bottom of the beam creates strain in the material. This strain is proportional to the distance from the Neutral Axis. According to Hooke's law, in a beam where the Modulus of Elasticity is constant across the section, the strain in the beam fibers is proportional to created flexure stress. Flexure stress in beams can be computed using the following equation:

$$\sigma = \frac{Mc}{I}$$

Where:
M= Bending moment
I= 2nd Moment of Area
c = h/2 at extreme fibers of a symmetric section.

The above equation can be re-stated by replacing c/I with 1/S. Where S is the section modulus of the beam. Thus:

$$\sigma = \frac{M}{S}$$

Shear stress

Shear stress is created by a shear force distributed across the section of the beam. Shear stress can be longitudinal or transverse. Just like flexure stress, this distribution is not uniform across the section. Shear stress can be calculated by either simply dividing the applied load by the area of the cross-section of the beam or using the following equation:

$$f_v = \frac{VQ}{Ib}$$

Where:

f_V = shear stress capacity

V= maximum shear stress

Q = 1st moment of area

I = 2nd moment of area

b = width of the beam

Shear stress will be maximum at locations where:

- V is high, for example, at the supports of the beam
- Q is high, for example, at the neutral axis
- b is low, for example, where the web width is thin
- I is low, for example, in less stiff section

To stabilize beams against the shear stress, stirrups are included in reinforced concrete beams, or steel plates are bolted or welded to steel beams where shear stress is critical.

Review

The following image represents axial, bending, and shear stresses, as well as the corresponding equations for calculating the respective values.

Compression	$\sigma = \frac{P}{A}$	
Tension	$\sigma = \frac{P}{A}$	
Flexure	$\sigma = \frac{Mc}{I}$	
Shear	$\tau = \frac{P}{A}$ *or* $\frac{VQ}{Ib}$	

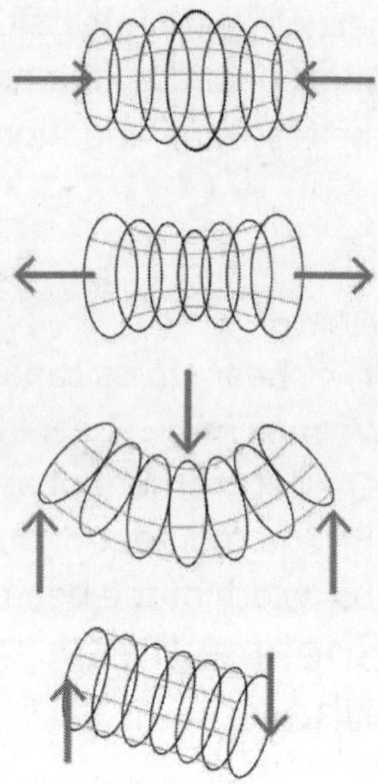

Figure 9-2: Corresponding equations for calculating axial, bending, and shear stresses

ALLOWABLE STRESS DESIGN METHOD

There are different methods of designing beams. Allowable Stress Design is a unique design practice that requires designers ensure that the stresses imposed on the structures don't exceed the elastic limit of the structural element. The allowable stress is determined by a factor of safety and the yield strength of the material. The allowable stress for different building materials is defined in building codes. For example, the allowable bending stress of structural steel is calculated by multiplying 0.66 and the steel yield stress. Likewise, the allowable bending stress of various species of structural wood is between 1000 to 600 psi.

You can design a beam by the following steps:

1. Choosing a steel grade and allowable stress.
2. Determining the bending moment either by solving a

free-body diagram or referring to design manuals or online beam calculators.

3. Calculate the section modulus (S_x) using the following equation:

$S_x = M/F_b$

Where:

S_x = section modulus = I_x/c ($c = h/2$ at extreme fibers of a symmetric section)

M = maximum bending (bending at the extreme fiber)

F_b = allowable bending stress (determined based on the material properties)

4. Choose a safe section with a suitable S_x from the tables provided in design manuals

BEAM DESIGN – EXAMPLE 1

A cantilever beam is loaded by a 0.4 k/ft uniformly distributed load along 8' of its length as shown in the image below.

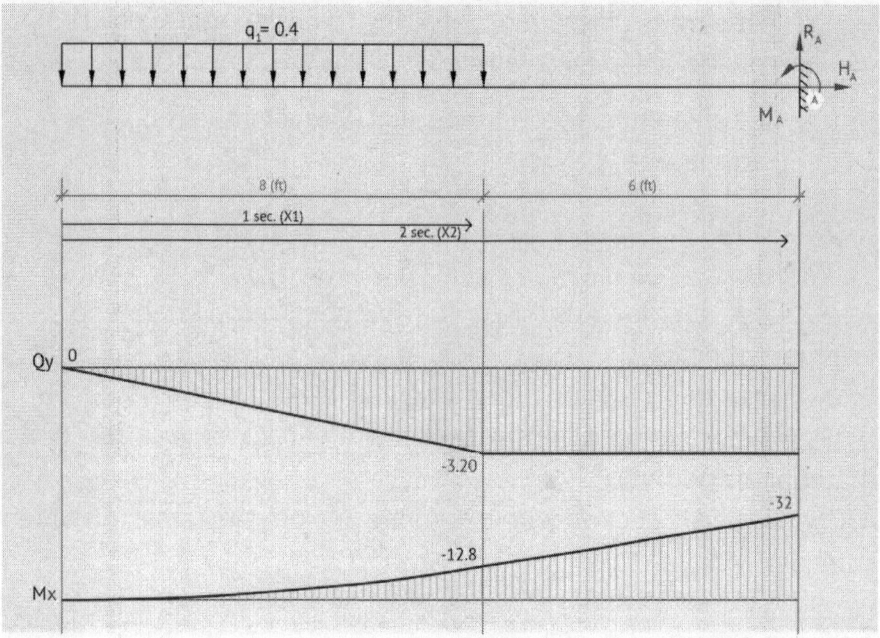

The beam end reaction, shear and bending diagrams are obtained using the the online beam calculator, BeamGuru. Find the suitable wide-flange steel or glulam timber section regarding the allowable bending stress limit.

Solving for a wide-flange steel beam:

First, we select a steel grade whose yield stress equals 50 ksi.

$F_y = 50$ ksi $\qquad F_b = 0.6\ F_y$

$S_x = M_{max}/F_b$

$S_x = (32000 \times 12)/(0.6 \times 50000) = 12.8\ in^3$

Referring to the steel construction manual, any wide-flange section whose $S_x \geq 12.8\ in^3$ will be suitable.

ALLOWABLE STRESS DESIGN SELECTION TABLE
For shapes used as beams S_x

F_y = 50 ksi			S_x	Shape	Depth d	F'_y	F_y = 36 ksi		
L_c	L_u	M_R					L_c	L_u	M_R
Ft	Ft	Kip-ft	In³		In	Ksi	Ft	Ft	Kip-ft
2.9	**3.6**	**47**	**17.1**	**W 12×16**	**12**	—	**4.1**	**4.3**	**34**
5.4	14.4	46	16.7	W 6×25	6⅜	—	6.4	20.0	33
3.6	4.4	45	16.2	W 10×17	10⅛	—	4.2	6.1	32
4.7	7.1	42	15.2	W 8×18	8⅛	—	5.5	9.9	30
2.5	**3.6**	**41**	**14.9**	**W 12×14**	**11⅞**	**54.3**	**3.5**	**4.2**	**30**
3.6	3.7	38	13.8	W 10×15	10	—	4.2	5.0	27
5.4	11.8	37	13.4	W 6×20	6¼	62.1	6.4	16.4	27
5.3	12.5	36	13.0	M 6×20	6	—	6.3	17.4	26
1.9	**2.6**	**33**	**12.0**	**M 12×11.8**	**12**	—	**2.7**	**3.0**	**24**
3.6	5.2	32	11.8	W 8×15	8⅛	—	4.2	7.2	23
2.8	3.6	30	10.9	W 10×12	9⅞	47.5	3.9	4.3	22

(Image source [15])

Solving for a glulam timber:

First, we select a glulam timber section with an 8 ¾" width and allowable stress of 1250 psi (DF grade L3). Note that the width of the beam is usually determined based on the dimensions of the columns that hold the beam.

$S_x = M_{max}/F_b$

$S_x = (32000 \times 12)/1250 = 307.2\ in^3$

Table 1C **Section Properties of *Western Species* Structural Glued Laminated Timber (Cont.)**

Depth	Area	X-X Axis			Y-Y Axis	
d (in.)	A (in.²)	I_x (in.⁴)	S_x (in.³)	r_x (in.)	I_y (in.⁴)	S_y (in.³)
		8-3/4 in. Width			**(r_y = 2.526 in.)**	
9	78.75	531.6	118.1	2.598	502.4	114.8
10-1/2	91.88	844.1	160.8	3.031	586.2	134.0
12	105.0	1260	210.0	3.464	669.9	153.1
13-1/2	118.1	1794	265.8	3.897	753.7	172.3
15	131.3	2461	328.1	4.330	837.4	191.4
16-1/2	144.4	3276	397.0	4.763	921.1	210.5
18	157.5	4253	472.5	5.196	1005	229.7
19-1/2	170.6	5407	554.5	5.629	1089	248.8
21	183.8	6753	643.1	6.062	1172	268.0

(Image source [16])

CHAPTER 10

Deflection in Simple Beams

Building structures are expected to be designed for strength, stability, and serviceability. Therefore, when designing a structural system with suitable strength, designers should consider tension rupture, compression crushing, and flexure in structural components. The stability of structural systems should be addressed by controlling buckling in columns and lateral-torsional buckling in beams. Finally, a structural system is expected to be not only a safe construction but also provide the building occupants a sense of comfort. Moreover, the structural system should not impact the durability of other building assemblies, such as interior finishing. Thus, beam deflection, building story drift, and cracking should be prevented in structures. In this chapter, the deflection of beams is discussed.

DEFINITION AND PRINCIPLES

In a simple beam, the compressive and tensile forces above and below the neutral axis result in a shortening and lengthening of the longitudinal fibers respectively above and below the neutral axis. This effect causes bending in the beam and displaces the beam fibers from their original positions. By definition, deflection is the displacement of the beam from its original horizontal position when subjected to loads. Various guidelines have been derived to determine maximum allowable deflection limits.

Typically, a floor system with a live load deflection in excess of L/360 will feel bouncy or crack plaster. The maximum deflection in a simple beam under a point load can be calculated using the following equation:

$$\Delta = \frac{PL^3}{48EI}.$$

Where:

P = the magnitude of the point load in lbs or kips
L = length of the beam (usually in ft)
E = Young's Modulus of the material
I = 2nd moment of area of the beam

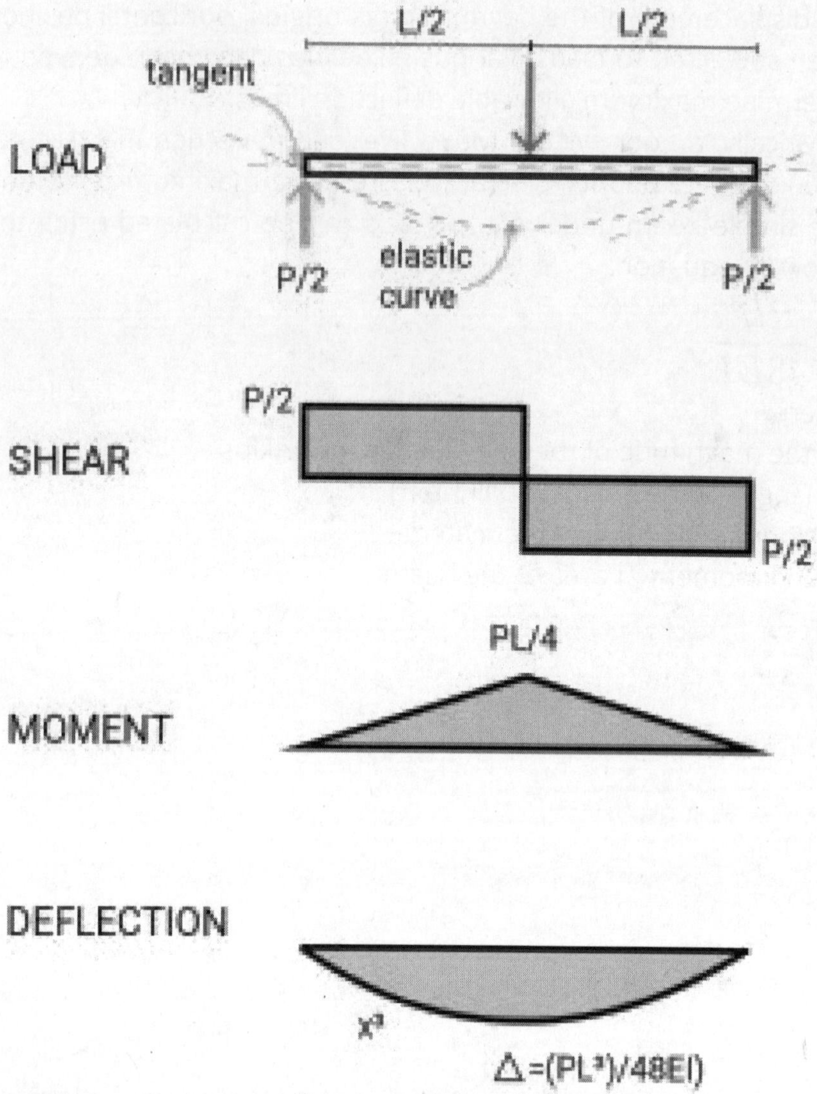

Figure 10-1: Maximum deflection, shear and bending stresses, and end reactions in a simple beam under a point load

The maximum deflection in a simple beam under a uniformly distributed load can be calculated using the following equation:

$$\Delta = \frac{5wL^4}{384EI}.$$

Where:

w = the magnitude of the distributed load in linear foot
L = length of the beam (usually in ft)
E = Young's Modulus of the material
I = 2nd moment of area of the beam

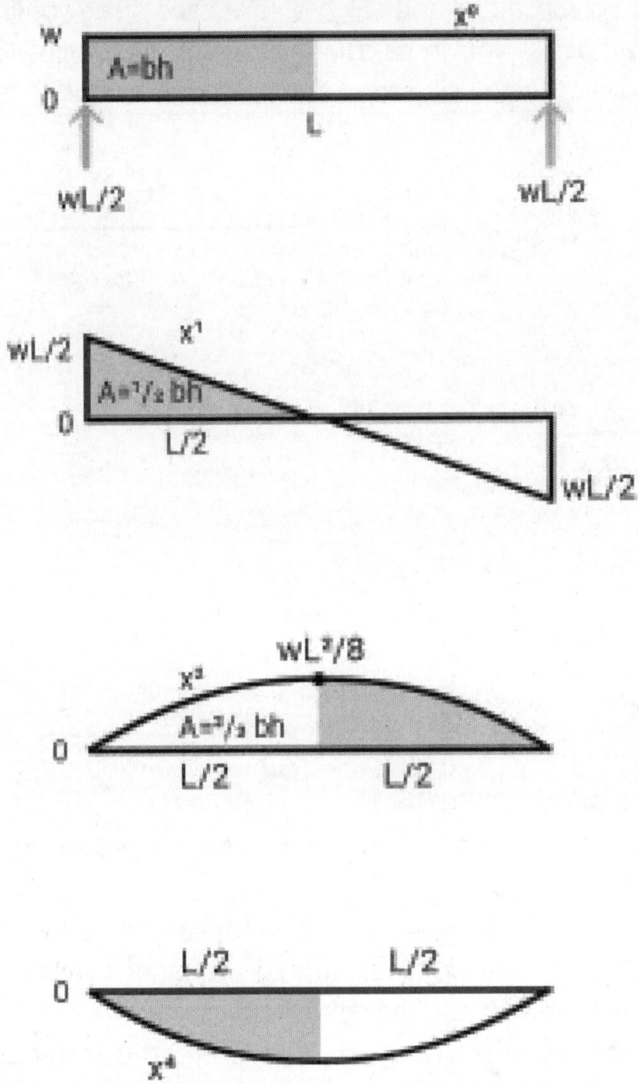

Figure 10-2: *Maximum deflection, shear and bending stresses, and end reactions in a simple beam under a uniformly distributed load*

The maximum deflection in different types of beams can be

obtained by drawing a free-body diagram or referring to the AISC Steel Construction Manual and using the provided tables showing the shear, bending, and deflection diagrams. Furthermore, online free beam calculators can be used to get maximum shear and bending values. ClearCalcs, and SkySiv are two free online platforms that can assist you in calculating beams. The following image shows a solved beam using the ClearCalcs online tool.

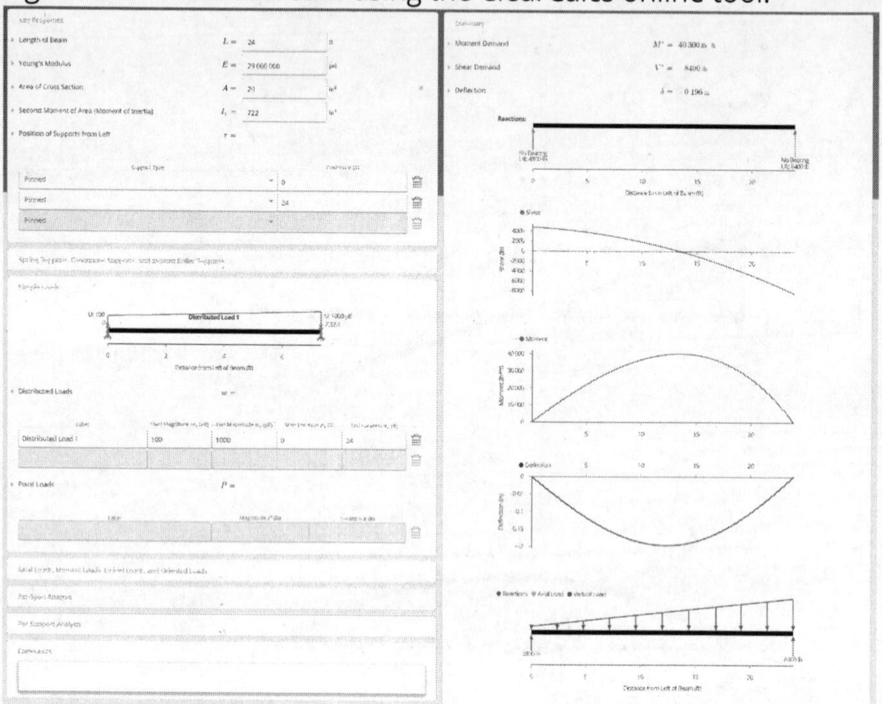

Figure 10-3: Diagrams of shear stress, bending stress and deflection of a beam solved by the ClearCalcs online tool

By "superposition," equations can be added for combination load cases. Care should be taken that added equations all give deflection at the same point, e.g. the centerline.

Note that if beam lengths and load (w) are entered in feet, a conversion factor of 1728 in3/ft3 must be applied in order to compute deflection in inches.

UNDERSTANDING THE RELATIONSHIP BETWEEN THE MAXIMUMS

A series of relationships between forces and deformations along a beam can be helpful in analysis. Using either the deflection or load as a starting point, the following characteristics can be discovered in a simple beam under a point load applied to the mid-span:

- Maximum slope occurs at the ends of the beam
- The point of zero slope occurs at the centerline. This is the point of maximum deflection
- The moment is positive for gravity loads.
- Shear and slope have balanced + and – areas.
- Deflection is negative for gravity loads.

The following characteristics can be discovered in a cantilever beam under a point load applied to its free end:

- The fixed end has the maximum moment but has a zero slope and deflection
- The free end has the maximum slope and deflection but a zero moment.

THE IDEA OF PRE AND POST-STRESSING

One of the methods of making the reinforced concrete beams more efficien is using pre-stressing or post-tensioning, where a beam is permanently loaded in a way that stresses are built up in the member opposite to those developed by the external loads.

Post tensioning

For constructing a post-tensioned concrete beam, a framework is positioned, hollow sheaths containing unstressed cables are draped into place, and concrete is cast around the sheaths. After the concrete is cured, cables are tensioned by jacks at each end of the beam. When the framework is removed, the cable force is maintained by permanent anchors at each end.

One or more interactive elements has been excluded from this version of the text. You can view them online here: https://pdx.pressbooks.pub/archistructures/?p=38#oembed-1

Video 10-1: Post-tensioning system (https://www.youtube.com/watch?time_continue=1&v=eQ2fJEbvJBs&feature=emb_logo)

Pre-tensioning

In a pre-stressed concrete beam, steel cables are pre-tensioned between abutments using hydraulic jacks. Then, concrete is cast around pre-tensioned cables and allowed to cure. After the concrete is cured, cables are cut. Cables apply a compressive force to the ends of the beam at the bottom level. This causes the beam to bow up, and the created curve offsets the deflection once the beam is loaded.

One or more interactive elements has been excluded from this version of the text. You can view them online here: https://pdx.pressbooks.pub/archistructures/?p=38#oembed-2

Video 10-2: Pre-stressing a reinforced concrete slab (https://www.youtube.com/watch?time_continue=1&v=0z6gjjrSn0M&feature=emb_logo)

TOPICS FOR CRITICAL THINKING

- Consider the location of the maximum deflection i beams, and explain which of the following beams is designed better?

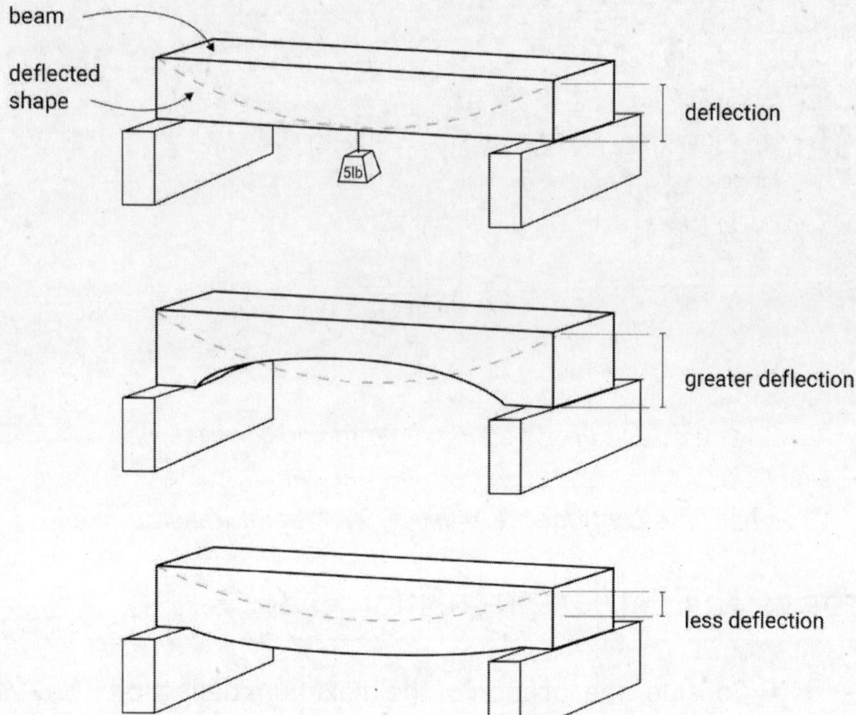

Figure 10-4: Deflection in beams with different longitudinal shapes

CHAPTER 11

Buckling in Columns

Columns' behavior in structural systems can be studied regarding their geometrical properties, loading type, and construction material. Following our discussion from the previous chapters, we will discuss the effect of columns' geometrical properties on potential failure modes of columns in buildings. Columns can be categorized into two contrasting groups of long and short columns. When the ratio of the effective length of a column to its least lateral dimension is equal to or less than 12, the column is considered short. In general, short columns fail by crushing.

$$f_c = \frac{P}{A} \leq F_c$$

Where:

f_c = Actual compressive stress

A = Cross-sectional area of column (in^2)

P = Load on the column

F_c = Allowable compressive stress per codes

Long or slender columns are those whose ratio of effective length to its least lateral dimension is more than 12. A long column fails due to buckling where the structural component under compression bows quickly. the buckling phenomenon in ling columns is due to the inevitable eccentricities in loading and the likelihood of irregularities in materials' resistance to compression. Long columns generally fail by buckling before reaching their limit of compressive stress. Accordingly, the load-carrying capacity of long columns is less than short columns.

One or more interactive elements has been excluded from this version of the text. You can view them online here: https://pdx.pressbooks.pub/archistructures/?p=40#oembed-2

Video 11-1: Demonstration of buckling in long elements in compression (https://www.youtube.com/watch?v=KjgOGw6SqIA&t=2s)

In 1757, a mathematician called Leonhard Euler derived a formula that gives the maximum axial load that a long slender column can carry without buckling. The Euler equation is given below:

$$F = \frac{\pi^2 IE}{KL^2}$$

Where:

F = Maximum allowable or critical compressive stress

A = Cross-sectional area (in^2)

E = Modulus of elasticity of the column material (psi)

K = Stiffness (curvature mode) factor; The K factor can be derived from the following table.

L = Column length between ends (inches)

Table 11-1: K factor in columns (source: column effect length, 2006, https://commons.wikimedia.org/wiki/File:ColumnEffectiveLength.png)

Buckled shape of column shown by dashed line	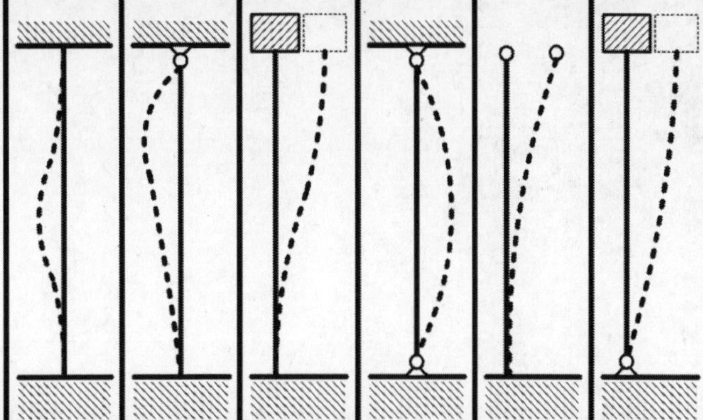					
Theoretical K value	0.5	0.7	1.0	1.0	2.0	2.0
Recommended design value K	0.65	0.80	1.2	1.0	2.10	2.0
End condition key		Rotation fixed and translation fixed Rotation free and translation fixed Rotation fixed and translation free Rotation free and translation free				

According to Leonhard Euler, the load under which a compressive element (such as a column) may buckle depends on its slenderness. The slenderness of a compressive element depends on:

- The length of the element
- The cross-sectional area
- 2nd moment of area
- Young's modulus of the element's material
- The support condition (K factor).

One or more interactive elements has been excluded from this version of the text. You can view them online here: https://pdx.pressbooks.pub/archistructures/?p=40#oembed-1

Video 11-2: How a column behaves under a compression force (https://www.youtube.com/watch?v=-ONVGO-iU4g&t=1s)

You may prevent buckling of the columns by:

- Decreasing the column length by bracing.
- Increasing the 2nd moment of area of the column by doubling the cross-sections or using box shapes.

TOPICS FOR CRITICAL THINKING

- Which of the following cross-sectional shapes will give a more stable column? Why?

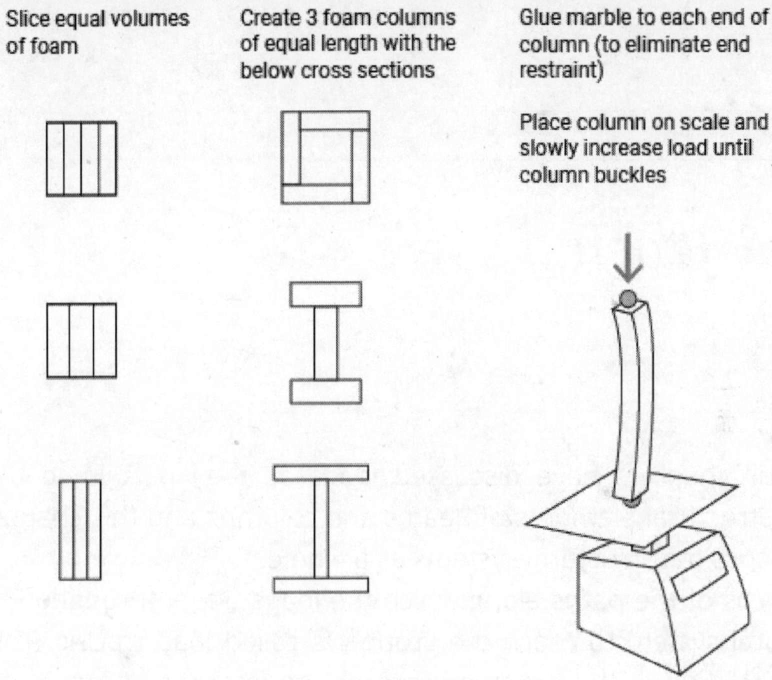

Figure 11-2: The effect of a column's cross-sectional shape on its stiffness

CHAPTER 12

Load Tracing

Previous chapters have discussed loads as they pertain to the major structural elements of beams and columns and this chapter will discuss the structural systems as a whole.

Analysis of the paths along which the loads travel throughout a structural system to reach the ground is called load tracing. This chapter discusses the systematic process of determining loads and support reactions of structural members as they affect the loading condition of other structural elements.

LOAD PATH

Structural systems are designed to transfer loads top down to the foundation and the ground. Loads travel throughout the structural components from the uppermost roof elements and move downward to the foundation system. This hierarchy can be described in the following order:

1. Flooring spans between joists
2. Joists span between beams
3. Beams span between girders
4. Girders span between columns

5. Columns carry load to the foundation system and then to the ground

Every time the load travels from one element to another, a support condition should be determined, and the loads and reactions at each transfer should be analyzed.

TRIBUTARY AREA

Each member is associated with an area that can be used to find the total load on that member. The tributary area is an area that corresponds to the load on a member. The following image shows a deck under a uniformly distributed live load. The deck is supported in three different ways: by bearing walls, joists and load-bearing walls, and a combination of joists, beams, girders, and columns. In each case, you can see the tributary area of the supporting walls and joists.

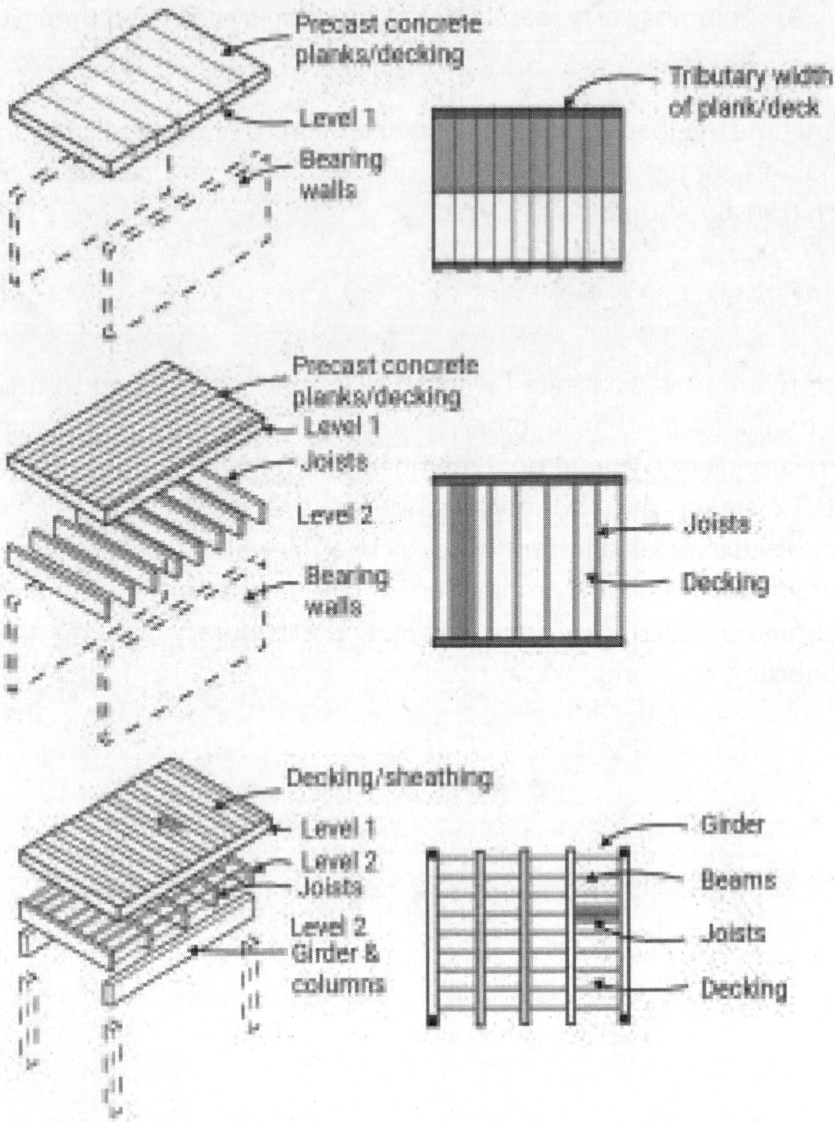

Figure 12-1: Tributary area of beams and joists in a simple deck system

The tributary area of one-way concrete slabs is calculated by splitting the slab area across the direction of steel reinforcements. In two-way concrete slabs, the tributary area divides at 45° from the corners.

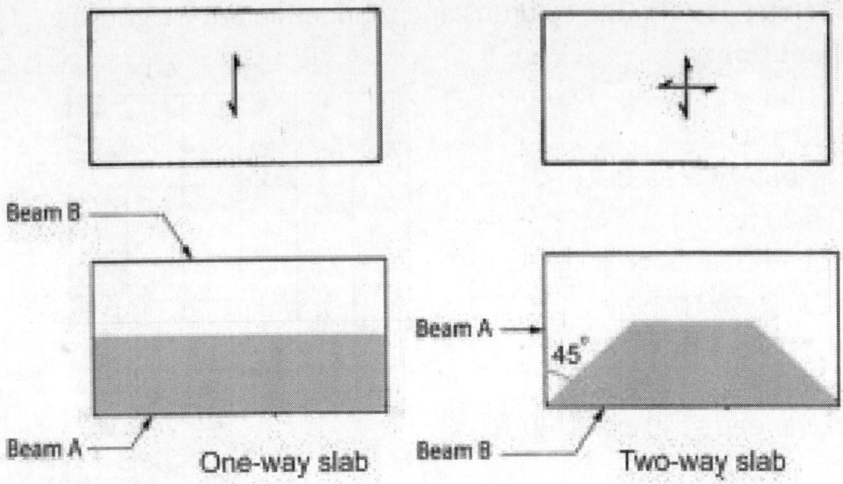

Figure 12-2: The tributary area of a one-way concrete slab vs. a two-way concrete slab

Example 1

The image below shows a one-way roof system with an area of 24×20 ft^2 . The roof system is composed of two girders, seven beams, and a 4-in thick concrete deck.

First, find the tributary area of Beam-1 and Beam-2.

Second, if:

Floor Live load = 40 psf

The weight of each beam is 31 plf

Weight of 1 cubic foot of the concrete = 150 lbs,

what type of and how much load do Beam-1, Beam-2, Girder-1 carry?

Third, four timber columns hold the roof system, and the allowable stress of the timber is 5000 psf, find the required cross-sectional area of a column using the following equation:

Allowable stress = P/A

Fourth, if the soil load-bearing capacity is 2000 psf, find the required surface area of the footing.

*Note: Assume the columns and foundation are weightless.

Solution:

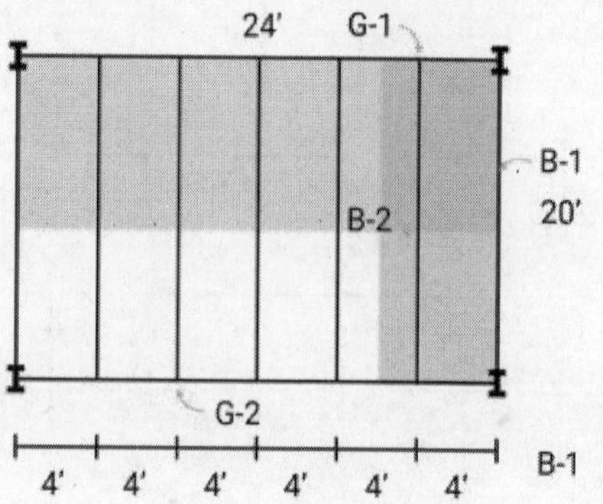

Tributary area Beam-1 = 2×20 = 40 ft^2

Tributary area Beam-2 = 4 × 20 = 80 ft^2

Dead load of deck = 150 ×(4″)/12 = 50 psf

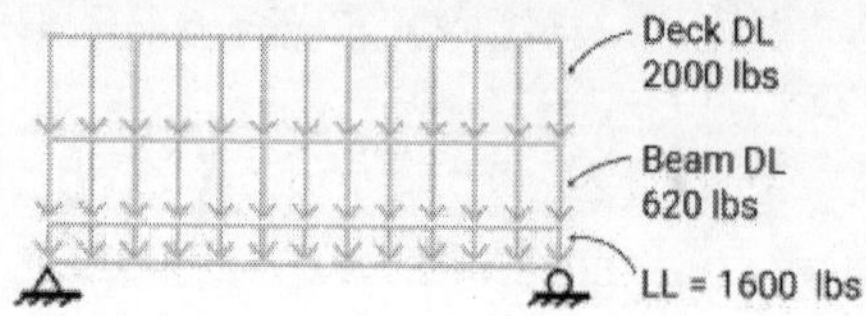

Beam-1

Live load = 40 psf × 40 ft^2 = 1600 lbs

DL of beam weight = 31 plf × 20 = 620 lbs

DL of deck = 50 psf × 40 ft^2= 2000 lbs

Beam-2

Live load = 40 psf × 80 ft^2 = 3200 lbs

DL of beam weight = 31 plf × 20 = 620 lbs

DL of deck = 50 psf × 80 ft^2 = 4000 lbs

Girder -1

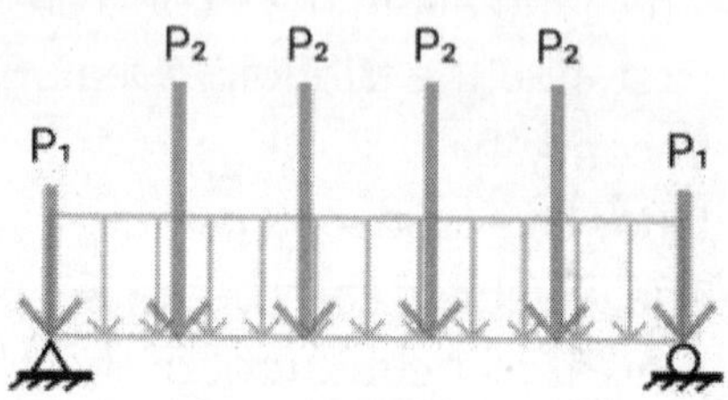

DL of beam weight = 31 plf × 24 = 744 lbs

Live load and DL of the deck are applied to Girder-1 via point loads P1 and P2.

P1 = end reaction in Beam-1 = (Total load on Beam-1)/2

= (1600 + 620 +2000)/2 = 4220 lbs/2 = 2110 lbs

P2 = end reaction in Beam-2 = (Total load on Beam-2)/2

= (3200 + 620 +4000)/2 = 7820 lbs/2 = 3910 lbs

Columns

Axial load on a column = (P1×2 + P2×5 + 744)/2 = (2110×2 + 3910 ×5 + 744)/2 = 24514/2 = 12257 lbs

Required cross sectional area A = P/ Allowable stress

A= 12257/5000 = 2.45 ft^2

Footing

Soil load bearing capacity = 2000 psf

Footing required cross sectional area A = P/ Allowable stress

A = 12257 / 2000 = 6.12 ft^2

Example 2

A concrete slab floor system spans in the directions shown. Find load diagrams for B1, B2, B3, B4, G2, G1.

The dead load of the floor slab is 70 psf.

The floor live load is 90 psf.

First, find the tributary area for each beam.

Second, calculate the distribution of the load on the floor.

Third, draw the load diagram for each beam.

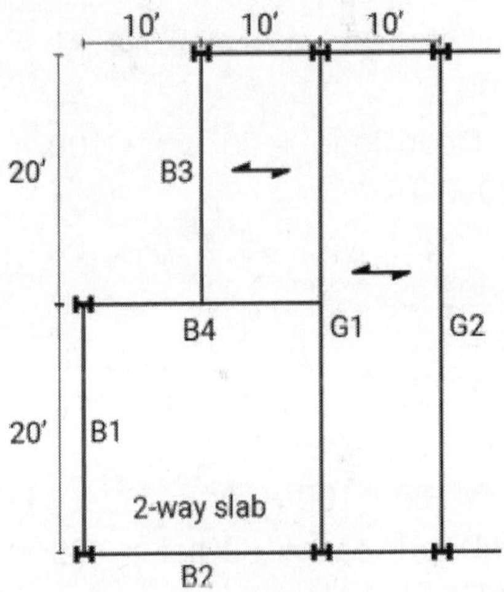

Solution:

B1

Tributary area = 20×10/2 = 100 ft^2

Dead load = 70 psf × 100 = 7000 lbs

Floor Live load = 90 × 100 = 9000 lbs

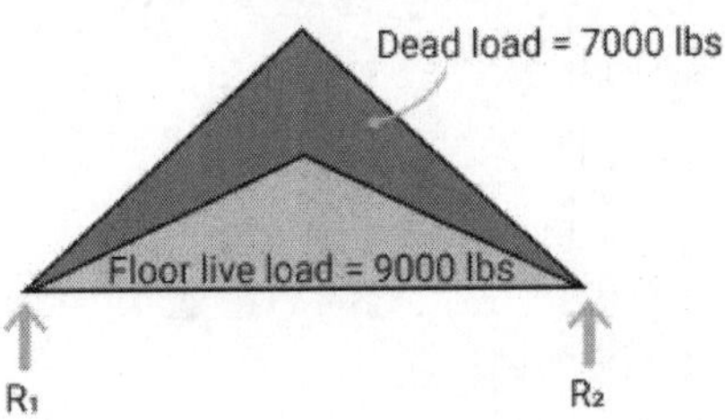

B2

Tributary area = 20×10/2 = 100 ft^2

Dead load = 70 psf × 100 = 7000 lbs

Floor Live load = 90 × 100 = 9000 lbs

B3

Tributary area = 20×10/2 = 100 ft^2

Dead load = 70 psf × 100 = 7000 lbs

Floor Live load = 90 × 100 = 9000 lbs

R1+R2-7000-9000= 0

R1+R2=16000

R1=R2 (symmetrical loading)

R1=R2 = 8000

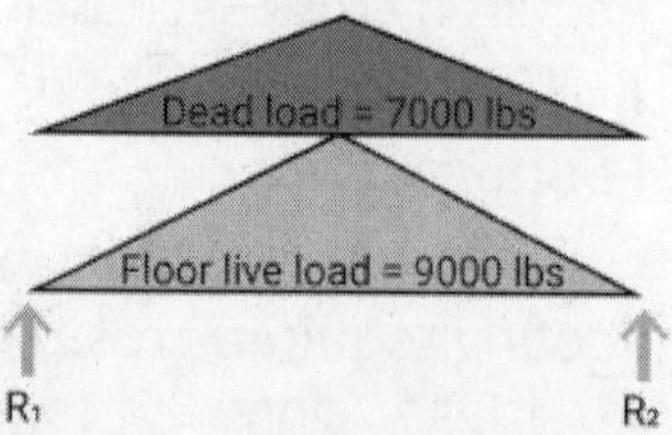

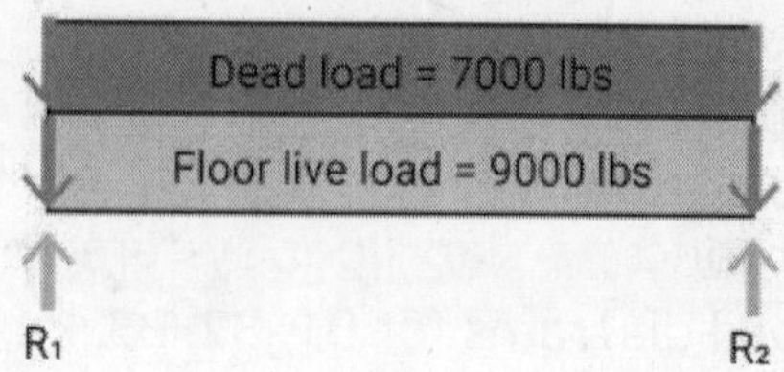

B4

Tributary area = 20×10/2 = 100 ft^2

Dead load = 70 psf × 100 = 7000 lbs

Floor Live load = 90 × 100 = 9000 lbs

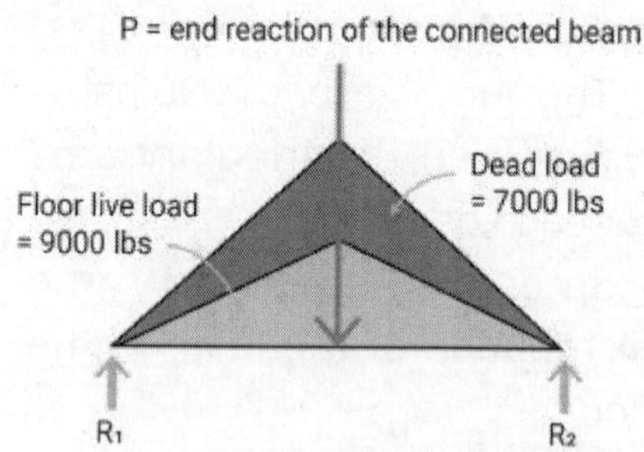

G1

Tributary area = 40×10/2 + 20×10/2 + 20×10/2

= 200 + 100 + 100 = 400 ft^2

Dead load = 70 psf × 400 = 28000 lbs

Floor Live load = 90 × 400 = 36000 lbs

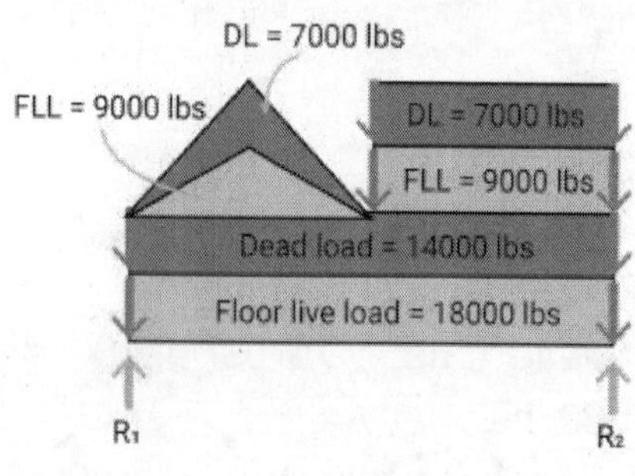

TOPICS FOR CRITICAL THINKING

What are the lateral force-resisting systems? Which of the three main kinds of a braced frame, a shear wall, and a moment-resisting frame is stiffer? Watch the following video (https://www.youtube.com/watch?v=kaKUHU3YYzY&t=1s) to learn about these systems.

One or more interactive elements has been excluded from this version of the text. You can view them online here: https://pdx.pressbooks.pub/archistructures/?p=42#oembed-1

Video 12-1: Lateral force-resisting systems

PART II

BIBLIOGRAPHY

[1] F. D. K. Ching, European building construction illustrated, Hoboken, New Jersey: John Wiley & Sons, 2014.

[2] B. Onouye and K. Kane, Statics and Strength of Materials for Architecture and Building Construction, 4th ed., Pearson, 2011.

[3] E. Allen and J. Iano, The Architect's Studio Companion: Rules of Thumb for Preliminary Design, 6th ed., Wiley, 2017.

[4] H. Nooshin, "Space structures and configuration processing," *Progress in Structural Engineering and Materials,* vol. 1, no. 3, pp. 329-336, 1998

[5] H. Nooshin and P. Disney, "Formex Configuration Processing I," *International Journal of Space Structures,* vol. 15, no. 1, pp. 1-52, 2000.

[6] H. Nooshin and P. Disney, "Formex Configuration Processing II," *in International Journal of Space Structures,* vol. 16, no. 1, 2001.

[7] H. Nooshin and P. Disney, "Formex Configuration Processing III," *International Journal of Space Structures,* vol. 17, no. 1, pp. 1-50, 2002.

[8] H. Nooshin, R. Kamyab and O. A. Samavati, "Exploring scallop forms," *International Journal of Space Structures,* vol. 32, no. 2, pp. 84-111, 2017.

[9] J.-F. Gabriel, Beyond the Cube: The Architecture of Space Frames and Polyhedra, Wiley, 1997.

[10] P. W. McMullin and J. S. Price, Introduction to Structures, Routledge, 2016.

[11] F. Moore, Understanding structures, New York: McGraw Hill Inc, 1998.

[12] M. Salvadori, Why Buildings Stand Up: The Strength of Architecture, New York: W. W. Norton & Company, 2002.

[13] J. Chilton, Space Grid Structures, Routledge, 2000

[14] D. Schodek and M. Bechthold, St ructures, 7th ed., Pearson, 2013.

[15] Steel Construction Manual, 4th ed., American Institute of Steel Construction, 2011

[16] National Design Specification: Design Values for Wood Construction, American Wood Council, 2015

[17] M. Moghimi, "Formex Configuration Processing of Compound and Freeform Structures," PhD dissertation, University of Surrey, Surrey, UK, 2006.

[18] A. Khodadadi, H. Nooshin, Z. Bozorgmehri, and M. Golabchi," Modern Lattice Domes Based on the Traditional Iranian Masonry Domes," *International Journal of Space Structures Annual Symposium*, vol 27, no 4, 2012, pp 231-245, ISSN: 0266-3511.

[19] H. Nooshin, "The Gallery," in *International Association for Shell and Spatial Structures Annual Symposium*, Boston, USA, July 2018.

[20] P. von Buelow and A. Khodadadi, "Computational form exploration of branching columns using concepts of formex algebra and the ParaGen method," in *International Association of Shells and Spatial Structures Annual Symposium 2020/21 and 7th International Conference on Spatial Structures*, Surrey, UK, August 2021.

[21] "Dr. Structure," YouTube, 18 September 2013. [Online]. Available: https://www.youtube.com/c/DrStructure/about. [Accessed 20 August 2021].

[22] "DartmouthX – The Engineering of Structures Around Us,"

YouTube, 27 March 2015. [Online]. Available: https://www.youtube.com/channel/UCTzQ-ZNy1DrKhchVBmPwU-Q/featured. [Accessed 20 August 2021].

[23] "Engineering Models," YouTube, 21 April 2015. [Online]. Available: https://www.youtube.com/channel/UCzgviV8WkULNua94BJDqI7g/about. [Accessed 20 August 2021].

[24] "Seeing and Touching Structural Concepts," University of Manchester, 2018. [Online]. Available: http://epsassets.manchester.ac.uk/structural-concepts/home. [Accessed 20 August 2021].

[25] "TED-Ed," YouTube, 1 March 2011. [Online]. Available: https://www.youtube.com/teded/featured. [Accessed 20 August 2021].

[26] "Formexia," [Online]. Available: http://formexia.com/. [Accessed 20 August 2021].

[27] "PhET," [Online]. Available: https://phet.colorado.edu/sims/html/vector-addition/latest/vector-addition_en.html. [Accessed 20 August 2021].

[28] "The Physics Classroom," [Online]. Available: https://www.physicsclassroom.com/Physics-Interactives/Vectors-and-Projectiles/Vector-Guessing-Game/Vector-Guessing-Game-Interactive. [Accessed 20 August 2021].

[29] "Cloud-based Structural Analysis Platform and Design Software Online for Calculating Beams, Frames, Trusses," [Online]. Available: https://beamguru.com/online/beam-calculator. [Accessed 20 August 2021].

[30] "Engineering Around Us: Beam Simulator," [Online]. Available: https://engsx.thayer.dartmouth.edu/simulations/#/beam. [Accessed 20 August 2021].

[31] "Sky Civ Cloud Engineering Software," [Online]. Available: https://skyciv.com/free-beam-calculator/. [Accessed 20 August 2021].

Index